Springer Theses

Recognizing Outstanding Ph.D. Research

Aims and Scope

The series "Springer Theses" brings together a selection of the very best Ph.D. theses from around the world and across the physical sciences. Nominated and endorsed by two recognized specialists, each published volume has been selected for its scientific excellence and the high impact of its contents for the pertinent field of research. For greater accessibility to non-specialists, the published versions include an extended introduction, as well as a foreword by the student's supervisor explaining the special relevance of the work for the field. As a whole, the series will provide a valuable resource both for newcomers to the research fields described, and for other scientists seeking detailed background information on special questions. Finally, it provides an accredited documentation of the valuable contributions made by today's younger generation of scientists.

Theses are accepted into the series by invited nomination only and must fulfill all of the following criteria

- They must be written in good English.
- The topic should fall within the confines of Chemistry, Physics, Earth Sciences, Engineering and related interdisciplinary fields such as Materials, Nanoscience, Chemical Engineering, Complex Systems and Biophysics.
- The work reported in the thesis must represent a significant scientific advance.
- If the thesis includes previously published material, permission to reproduce this must be gained from the respective copyright holder.
- They must have been examined and passed during the 12 months prior to nomination.
- Each thesis should include a foreword by the supervisor outlining the significance of its content.
- The theses should have a clearly defined structure including an introduction accessible to scientists not expert in that particular field.

More information about this series at http://www.springer.com/series/8790

Chang-Ki Moon

Molecular Orientation and Emission Characteristics of Ir Complexes and Exciplex in Organic Thin Films

Doctoral Thesis accepted by
Seoul National University, Seoul, Korea
(Republic of)

Springer

Author
Dr. Chang-Ki Moon
Department of Materials Science
and Engineering, The Graduate School
Seoul National University
Seoul, Korea (Republic of)

Supervisor
Prof. Jang-Joo Kim
Department of Materials Science
and Engineering
Seoul National University
Seoul, Korea (Republic of)

ISSN 2190-5053 ISSN 2190-5061 (electronic)
Springer Theses
ISBN 978-981-13-6054-1 ISBN 978-981-13-6055-8 (eBook)
https://doi.org/10.1007/978-981-13-6055-8

Library of Congress Control Number: 2018967419

This Springer imprint is published by the registered company Springer Nature Singapore Pte Ltd.
The registered company address is: 152 Beach Road, #21-01/04 Gateway East, Singapore 189721, Singapore

Supervisor's Foreword

Molecular orientation and electronic structure of organic semiconductors are important factors influencing the electrical and optical properties of the materials. This thesis describes Dr. Moon's fundamental work during his Ph.D. in the Department of Materials Science and Engineering at Seoul National University to understand the origin of the preferred orientation of doped molecules and electronic structure of intermolecular charge transfer complex, which are important in organic electronics. Simulations and calculations along with experiments provide clues to the analysis of molecular orientation and film properties. Dr. Moon has pioneered the analysis of the emission properties of solids according to the molecular orientation using molecular dynamics simulations and quantum chemical calculations. This thesis describes Dr. Moon's works on the molecular orientation and resulting film properties. Dr. Moon's first research interest in this area was to develop the optical simulation of luminescence from thin films and organic light-emitting diodes (OLEDs). Birefringence resulted from molecular orientations made it difficult to analyze thin-film luminescence. He established an optical model of the dipole radiation in an anisotropic medium and proved the model by experiments. It turns out that the model describes the emission characteristics of OLEDs very well. The same model was also utilized to analyze the emitting dipole orientation of iridium complexes, the most important emitting dyes used for OLEDs, to get accurate orientation of the emitting dipoles. It was known that many of iridium complexes used in OLEDs, even though they have globular shape, have preferred emitting dipole orientations as doped in organic amorphous layers, which is beneficial for OLEDs efficiency, but the reason of the orientation was ambiguous in those days. Dr. Moon simulated vacuum deposition process using molecular dynamics to understand the origin of the orientation of iridium complexes doped in organic layers and to predict the emitting dipole orientation. His experimental and calculation results improved the understanding of the alignment and distribution of iridium complexes and showed that intermolecular interactions between hosts and dopant molecules play a pivotal role to induce preferred orientation of dopant molecules during the vacuum process.

Dr. Moon combined the simulation method and orientation of emitting dipoles to understand the electronic structure of exciplex in solid. Exciplexes in solid states exhibit extraordinary characteristics including broad emission spectra, multi-exponential photoluminescence (PL) decay curves, and spectral redshifts as time delays in transient PL, whose origin was not fully understood at the time. Dr. Moon presented experimental and theoretical evidences that all of the emission characteristics of solid-state exciplexes originate from differences in their dimer configurations, which have different charge transfer rates, emission energies, singlet–triplet energy gaps, kinetic rate constants, and emitting dipole orientations. This conclusion is based on experimental observations, quantum chemical calculations, and molecular dynamics simulations. These results enabled us to develop a model of the electronic structure of an exciplex in a solid state. This comprehensive model accommodates all of the characteristics of the exciplex and can be used to further our understanding of solid-state charge transfer complexes playing important roles in OPVs and OLEDs.

This thesis covers the most advanced theories and analytical methods of the emission characteristics of doped organic films and electronic structures of intermolecular charge transfer complexes in solid in organic thin films. It is recommended for readers who desire to understand the orientation of molecules in organic amorphous films, its effects on optical properties of thin films as well as the electronic structure of intermolecular charge transfer complexes, which play important roles in organic electronics, especially in OLEDs and OPVs.

Seoul, Korea (Republic of) Prof. Jang-Joo Kim
November 2018

Abstract

Organic light-emitting diodes (OLEDs) have drawn great attention in lighting and display technologies with the advantages of high color purity, large-area processability, low cost, and flexibility. Development of the materials and the device structures is still demanded because the power efficiency of OLEDs is low to replace other lighting sources. The key importance of emitting materials for OLEDs is utilizing the non-radiative triplet excitons as photons and the increase of the outcoupling efficiency. The triplet excitons can be harvested as photons by employing phosphorescent or delayed fluorescent emitters. Control of the molecular orientation in organic thin films is one of the methods to enhance the outcoupling efficiency of OLEDs. Horizontal alignment of the emitting dipole moment to the substrate largely enhances the outcoupling efficiency of light in OLEDs.

The optical model of OLEDs interprets dipole radiation in a microcavity structure. It predicts external quantum efficiency and electroluminescence spectrum of OLEDs very precisely. In addition, the model has been recently applied to the method of determining the emitting dipole orientation (EDO) of an emissive layer. The preferred molecular orientations in organic thin films lead to optical birefringence so that the development of the optical model of OLEDs with consideration of the birefringence is required.

Ir(III) complex is one of the phosphorescent dyes for highly efficient OLEDs with high radiative quantum efficiency and various emission color spectrum. In general, Ir complexes have octahedral structures with three cyclometalating bidentate ligands and only small amount of them are doped in an organic host in the emissive layer. Therefore, their molecular orientations have been regarded as isotropic for a decade. However, the preferred horizontal orientation of the emitting dipole moment was investigated from a heteroleptic Ir complex in 2011. Optical simulation predicts over 46% of external quantum efficiency from the OLEDs using the Ir complexes with the perfect horizontal alignment of the emitting dipole moment. Now, the molecular orientation of Ir complex is an important topic of OLEDs.

Exciplex is a charge transfer complex formed by charge transfer between different molecules at the exciplex state. Exciplex has a very low singlet–triplet energy gap by spatial separation of the frontier orbitals, enabling delayed fluorescence with electron exchange between singlet and triplet states at the room temperature. Exciplex has been highlighted as an emitter of OLEDs because it also can harvest the non-radiative triplet state as delayed fluorescence. Many kinds of exciplexes have been reported with radiative quantum efficiency, but understanding the electronic structure and the emission mechanism of the exciplex is still insufficient. We need more concentration on the dimer arrangements and the charge transfer process in a solid-state blend to understand the emission mechanism of the exciplex.

This thesis analyzes three subjects: (1) modeling of dipole radiation from an anisotropic microcavity, (2) the origin of molecular orientation in organic thin films, and (3) electronic structure and emission process of exciplex in a solid state.

Chapter 1 introduces the molecular orientation in thin films, the optical modeling, the iridium complexes, and the exciplex related to OLEDs.

Chapter 2 describes optical modeling of the luminescence from an oriented emitting dipole moment in an anisotropic microcavity, and the model is expanded for optical analysis of OLEDs. The dipole radiation in an anisotropic microcavity is solved as functions of the dipole orientation and direction of the polarization using the classical dipole theory. As a result, the optical model enables to analyze the far-field radiation from an Ir complex doped in a birefringent layer. In addition, it successfully analyzed the angular emission emission spectra and external quantum efficiency of OLEDs employing the birefringent emissive layer.

Chapter 3 analyzes host effect on the orientation of the iridium complexes in organic amorphous film and furthers the intermolecular interactions between the Ir complexes and host molecules on organic surfaces during vacuum deposition via quantum chemical calculation and molecular dynamics simulations. Most researches have tried to control the molecular orientation by a change of the molecular structure of Ir complexes. However, this thesis investigates that the preferred orientation of the emitting dipole moment is able to vary from horizontal, isotropic, to rather vertical directions depending on host materials. The local attraction between the host–aromatic ligand of the dopant on the surface induces the horizontal EDO.

Furthermore, molecular dynamics (MD) simulations were demonstrated to simulate the vacuum deposition processes and observe the molecular configurations with various host–dopant combinations. As a result, the emitting dipole orientation is determined by the alignment of the aromatic ligands parallel to the substrate rather than the alignment of aliphatic ligands. Therefore, the host–dopant aromatic–aromatic interactions whose direction is parallel to the substrate is attributed to the origin of the preferred molecular orientation of heteroleptic iridium complexes in vacuum-deposited organic thin films. Energetic analysis indicated that the dispersion interaction is the major force for the orientation of iridium complexes, but the local electrostatic attraction induces further alignment if a polar host is used.

Chapter 4 analyzes the electronic structure and emission process of the exciplex in a solid-state blend employing the hybridized local and charge transfer excited state. The exciplex energy spectrum is calculated by a product of the exciplex energy and density of the molecule as functions of distance between the hetero-dimer. As a result, a variety of properties such as singlet and triplet levels, orientation of the transition dipole moment, and fluorescent rate, depend on the donor–acceptor binding geometry in the solid. Superposition of the fast-decaying high-energy exciplex and slow-decaying low-energy exciplex is attributed to a reason of the spectral redshift as time delays of the solid-state exciplex.

Keywords Organic light-emitting materials · Organic light-emitting diodes · Molecular orientation · Emitting dipole orientation · Vacuum deposition · Iridium complex · Exciplex · Optical modeling · Quantum chemical calculation · Molecular dynamics simulation

List of Publications

1st Author

1. Chang-Ki Moon, Jin-Suk Huh, Jae-Min Kim, and Jang-Joo Kim*, "Electronic structure and emission process of excited charge transfer states in solids", ***Chemistry of Materials***, 30, 5648–5654 (2018). A most downloaded paper on August 2018.
2. Chang-Ki Moon, Kwon-Hyeon Kim, and Jang-Joo Kim*, "Unraveling the orientation of phosphors doped in organic semiconducting layers", ***Nature Communications***, 8, 791 (2017).
3. Chang-Ki Moon, Katsuaki Suzuki, Katsuyuki Shizu, Chihaya Adachi, Hironori Kaji*, Jang-Joo Kim*, "Combined Inter- and Intramolecular Charge-Transfer Processes for Highly Efficient Fluorescent Organic Light-Emitting Diodes with Reduced Triplet Exciton Quenching", ***Advanced Materials***, 29, 1606448 (2017).
4. Chan Seok Oh(+), Chang-Ki Moon(+), Jeong Min Choi, Jin-Suk Huh, Jang-Joo Kim, Jun Yeob Lee*, "Relationship between molecular structure and dipole orientation of thermally activated delayed fluorescent emitters", ***Organic Electronics***, 42, 337–342 (2017). (+: equal contribution)
5. Chang-Ki Moon, Kwon-Hyeon Kim, Jin Woo Lee, and Jang-Joo Kim*, "Influence of host molecules on emitting dipole orientation of phosphorescent iridium complexes", ***Chemistry of Materials***, 27 (8), 2767–2769 (2015).
6. Chang-Ki Moon, Sei-Yong Kim, Jeong-Hwan Lee, and Jang-Joo Kim*, "Luminescence from oriented emitting dipoles in a birefringent medium" ***Optics Express***, 23, A279–A291 (2015).
7. Jeong-Hwan Lee(+), Ganguri Sarada(+), Chang-Ki Moon(+), Woosum Cho, Kwon-Hyeon Kim, Young Geun Park, Jin Yong Lee, Sung-Ho Jin*, and Jang-Joo Kim*, "Finely Tuned Blue Iridium Complexes with Varying Horizontal Emission Dipole Ratios and Quantum Yields for Phosphorescent Organic Light-Emitting Diodes", ***Advanced Optical Materials***, 3, 211–220 (2014).

Other Publications

1. Jinouk Song, Kwon-Hyeon Kim, Eunhye Kim, Chang-Ki Moon, Yun-Hi Kim, Jang-Joo Kim, and Seunghyup Yoo*, "Lensfree OLEDs with over 50% external quantum efficiency via external scattering and horizontally oriented emitters", ***Nature Communications**, In press.*
2. Jin-Suk Huh, Kwon-Hyeon Kim, Chang-Ki Moon, Jang-Joo Kim*, "Dependence of Pt(II) based phosphorescent emitter orientation on host molecule orientation in doped organic thin films", ***Organic Electronics***, 45, 279–284 (2017).

3. Hyun-Gu Kim, Kwon-Hyeon Kim, Chang-Ki Moon, Jang-Joo Kim*, "Harnessing Triplet Excited States by Fluorescent Dopant Utilizing Codoped Phosphorescent Dopant in Exciplex Host for Efficient Fluorescent Organic Light Emitting Diodes", ***Advanced Optical Materials***, 5 (3), 1600749 (2017).
4. Bomi Sim, Chang-Ki Moon, Kwon-Hyeon Kim, Jang-Joo Kim*, "Quantitative Analysis of the Efficiency of OLEDs", ***ACS Applied Materials & Interfaces***, 8 (48), 33010–33018 (2016).
5. Ganguri Sarada(+), Bomi Sim(+), Chang-Ki Moon, Woosum Cho, Kwon-Hyeon Kim, Vijaya Gopalan Sree, Eunju Park, Jang-Joo Kim*, Sung-Ho Jin*, "Synthesis and characterization of highly efficient blue Ir(III) complexes by tailoring β-diketonate ancillary ligand for highly efficient PhOLED applications", ***Organic Electronics***, 39, 91–99 (2016).
6. Kwon-Hyeon Kim(+), Jang Yeol Baek(+), Chan Woo Cheon, Chang-Ki Moon, Bomi Sim, Myeong Yong Choi, Jang-Joo Kim*, Yun-Hi Kim*, "Highly efficient non-doped deep blue fluorescent emitters with horizontal emitting dipoles using interconnecting units between chromophores", ***Chemical Communications***, 52 (73), 10956–10959 (2016).
7. Jae-Min Kim, Seung-Jun Yoo, Chang-Ki Moon, Bomi Sim, Jae-Hyun Lee*, Heeseon Lim, Jeong Won Kim, Jang-Joo Kim*, "N-type Molecular Doping in Organic Semiconductors: Formation and Dissociation Efficiencies of Charge Transfer Complex", ***The Journal of Physical Chemistry C***, 120 (17), 9475–9481 (2016).
8. Hyun Shin(+), Jeong-Hwan Lee(+), Chang-Ki Moon, Jin-Suk Huh, Bomi Sim, Jang-Joo Kim*, "Sky-Blue Phosphorescent OLEDs with 34.1% External Quantum Efficiency Using a Low Refractive Index Electron Transporting Layer", ***Advanced Materials***, 28 (24), 4920–4925 (2016).
9. Jin Won Sun, Kwon-Hyeon Kim, Chang-Ki Moon, Jeong-Hwan Lee, Jang-Joo Kim*, "Highly efficient sky-blue fluorescent organic light emitting diode based on mixed co-host system for thermally activated delayed fluorescence emitter (2CzPN)", ***ACS Applied Materials & Interfaces***, 8 (15), 9806–9810 (2016).
10. Kwon-Hyeon Kim, Jia-Ling Liao, Si Woo Lee, Bomi Sim, Chang-Ki Moon, Gene-Hsiang Lee, Hyo Jung Kim, Yun Chi*, Jang-Joo Kim*, "Crystal Organic Light-Emitting Diodes with Perfectly Oriented Non-Doped Pt-Based Emitting Layer", ***Advanced Materials***, 28 (13), 2526–2532 (2016).
11. Hyun-Sub Shim, Chang-Ki Moon, Jihun Kim, Chun-Kai Wang, Bomi Sim, Francis Lin, Ken-Tsung Wong, Yongsok Seo, Jang-Joo Kim*, "Efficient Vacuum-Deposited Ternary Organic Solar Cells with Broad Absorption, Energy Transfer, and Enhanced Hole Mobility", ***ACS Applied Materials & Interfaces***, 8 (2), 1214–1219 (2016).
12. Jin Won Sun(+), Jang Yeol Baek(+), Kwon-Hyeon Kim, Chang-Ki Moon, Jeong-Hwan Lee, Soon-Ki Kwon, Yun-Hi Kim*, Jang-Joo Kim*, "Thermally Activated Delayed Fluorescence from Azasiline Based Intramolecular Charge-Transfer Emitter (DTPDDA) and a Highly Efficient Blue Light Emitting Diode", ***Chemistry of Materials***, 27 (19), 6675–6681 (2015).

13. Hyun-Sub Shim, Francis Lin, Jihun Kim, Bomi Sim, Tae-Min Kim, Chang-Ki Moon, Chun-Kai Wang, Yongsok Seo*, Ken-Tsung Wong*, and Jang-Joo Kim*, "Efficient Vacuum-Deposited Tandem Organic Solar Cells with Fill Factors Higher Than Single-Junction Subcells", ***Advanced Energy Materials***, 5 (13), 1500228 (2015).
14. Kwon-Hyeon Kim(+), Jae-Yeol Ma(+), Chang-Ki Moon, Jeong-Hwan Lee, Jang Yeol Baek, Yun-Hi Kim*, and Jang-Joo Kim*, "Controlling Emitting Dipole Orientation with Methyl Substituents on Main Ligand of Iridium Complexes for Highly Efficient Phosphorescent Organic Light-Emitting Diodes", ***Advanced Optical Materials***, 3 (9), 1191–1196 (2015).
15. Sohee Jeon, Jeong-Hwan Lee, Jun-Ho Jeong, Young Seok Song, Chang-Ki Moon, Jang-Joo Kim, and Jae Ryoun Youn*, "Vacuum Nanohole Array Embedded Phosphorescent Organic Light Emitting Diodes" ***Scientific Reports***, 5, 8685 (2015).
16. Kwon-Hyeon Kim, Chang-Ki Moon, Jin Won Sun, Bomi Sim, and Jang-Joo Kim*, "Triplet Harvesting by a Conventional Fluorescent Emitter Using Reverse Intersystem Crossing of Host Triplet Exciplex", ***Advanced Optical Materials***, 3, 895–899 (2015).
17. Jung-Bum Kim, Jeong-Hwan Lee, Chang-Ki Moon, Kwon-Hyeon Kim, and Jang-Joo Kim*, "Highly enhanced light extraction from organic light emitting diodes with little image blurring and good color stability", ***Organic Electronics***, 17, 115–120 (2015).
18. Dae-Ho Kim, Kyu-Sik Kim, Hyun-Sub Shim, Chang-Ki Moon, Yong Wan Jin, and Jang-Joo Kim*, "A high performance semitransparent organic photodetector with green color selectivity", ***Applied Physics Letters***, 105, 213301 (2014).
19. Kwon-Hyeon Kim, Sunghun Lee, Chang-Ki Moon, Sei-Yong Kim, Young-Seo Park, Jeong-Hwan Lee, Jin Woo Lee, June Huh, Youngmin You, Jang-Joo Kim*, "Phosphorescent dye-based supramolecules for high-efficiency organic light-emitting diodes", ***Nature Communications***, 5, 4769 (2014).
20. Jung-Bum Kim, Jeong-Hwan Lee, Chang-Ki Moon, Kwon-Hyeon Kim, Jang-Joo Kim*, "Highly efficient inverted top emission organic light emitting diodes using a horizontally oriented green phosphorescent emitter", ***Organic Electronics***, 15, 2715–2718 (2014).
21. Thota Giridhar(+), Jeong-Hwan Lee(+), Woosum Cho, Hyunyoung Yoo, Chang-Ki Moon, Jang-Joo Kim*, Sung-Ho Jin*, "Highly efficient bluish green phosphorescent organic light-emitting diodes based on heteroleptic iridium(III) complexes with phenylpyridine main skeleton", ***Organic Electronics***, 15, 1687–1694 (2014).
22. Jin Won Sun(+), Jeong-Hwan Lee(+), Chang-Ki Moon, Kwon-Hyeon Kim, Hyun Shin and Jang-Joo Kim*, "A Fluorescent Organic Light Emitting Diode with 30% External Quantum Efficiency", ***Advanced Materials***, 26, 5684–5688 (2014).

23. Hyun Shin(+), Sunghun Lee(+), Kwon-Hyeon Kim, Chang-Ki Moon, Seung-Jun Yoo, Jeong-Hwan Lee and Jang-Joo Kim*, "Blue Phosphorescent Organic Light Emitting Diodes using an Exciplex Forming Co-host with the External Quantum Efficiency of Theoretical Limit", ***Advanced Materials***, 26, 4730 (2014).
24. Kwon-Hyeon Kim, Chang-Ki Moon, Jeong-Hwan Lee, Sei-Yong Kim and Jang-Joo Kim*, "Highly Efficient Organic Light-Emitting Diodes with Phosphorescent Emitters Having High Quantum Yield and Horizontal Orientation of Transition Dipole Moments", ***Advanced Materials***, 26, 3844 (2014).
25. Jung-Bum Kim(+), Jeong-Hwan Lee(+), Chang-Ki Moon, Jang-Joo Kim*, "Highly efficient inverted top emitting organic light emitting diodes using a transparent top electrode with color stability on viewing angle", ***Applied Physics Letters***, 104, 073301 (2014).
26. Seung-Jun Yoo, Jung-Hung Chang, Jeong-Hwan Lee, Chang-Ki Moon, Chih-I Wu, and Jang-Joo Kim*, "Formation of perfect ohmic contact at indium tin oxide/N,N9-di(naphthalene-1-yl)-N,N9-diphenyl-benzidine interface using ReO3", ***Scientific Reports***, 4, 3902 (2014).
27. Jung-Bum Kim, Jeong-Hwan Lee, Chang-Ki Moon, Sei-Yong Kim, Jang-Joo Kim "Highly Enhanced Light Extraction from Surface Plasmonic Loss Minimized Organic Light-Emitting Diodes", ***Advanced Materials***, 25, 3571–3577 (2013).
28. Sei-Yong Kim, Won-Ik Jeong, Christian Mayr, Yong-Seo Park, Kwon-Hyeon Kim, Jeong-Hwan Lee, Chang-Ki Moon, Wolfgang Brutting and Jang-Joo Kim, "Organic Light-Emitting Diodes with 30% External Quantum Efficiency Based on a Horizontally Oriented Emitter", ***Advanced Functional Materials***, 23, 3896 (2013).

List of Presentations

1. Chang-Ki Moon, Kwon-Hyeon Kim, Jang-Joo Kim, "Origin of the Orientation of Ir Complexes Doped in Organic Semiconducting Layers" MRS Spring 2018, April 2–6 (April 3), 2018, USA (Poster).
2. Chang-Ki Moon, Jin-Suk Huh, Jae-Min Kim, Jang-Joo Kim, "Electronic Structures and Emission Processes of Exciplex depending on Dimer Geometries in Solid States", The 17th International Meeting on Information Display (iMiD 2017), August 28–31 (August 31), 2017, Korea (Oral).
3. Chang-Ki Moon, Kwon-Hyeon Kim, Jang-Joo Kim, "Unraveling origin of phosphors doped in organic semiconducting layers", SPIE Optics+Photonics 2017, August 6–10 (August 8), 2017, USA (Oral).
4. Chang-Ki Moon, Jin-Suk Huh, Jae-Min Kim, Jang-Joo Kim, "Electronic structures and emission processes of exciplex depending on dimer

configurations", 2nd International TADF Workshop, July 19–21 (July 19), 2017, Japan (Poster).

5. Chang-Ki Moon, Kwon-Hyeon Kim, Jang-Joo Kim, "Unraveling the orientation of phosphors doped in organic semiconducting layers", The 28th International Conference on Molecular Electronics and Devices 2017 (IC ME&D 2017), May 18–19 (May 19), 2017, Korea (Poster).
6. Chang-Ki Moon, Katsuaki Suzuki, Katsuyuki Shizu, Chihaya Adachi, Hironori Kaji, Jang-Joo Kim, "Combination of an exciplex host and a TADF dopant for efficient fluorescent OLEDs with low roll-off", The 8th Asian Conference on Organic Electronics (A-COE 2016), December 5–7 (December 6), 2016, Japan (Poster).
7. Chang-Ki Moon, Katsuaki Suzuki, Katsuyuki Shizu, Hironori Kaji, Jang-Joo Kim, "Combined Inter- and Intra-molecular Charge Transfer Processes for Highly Efficient Fluorescent Organic Light-emitting Diodes with Reduced Triplet Exciton Quenching", International Conference on Electronic Materials and Nanotechnology for Green Environment (ENGE 2016), November 6–9 (November 07), 2016, Korea (Poster).
8. Chang-Ki Moon, Sei-Yong Kim, Jeong-Hwan Lee, Jang-Joo Kim, "Luminescence from oriented emitting dipoles in a birefringent medium", The International Symposium on Recent Advances and Future Issues in Organic Electroluminescence (ISOEL2016), February 17–19 (February 18), 2016, Korea (Poster).
9. Chang-Ki Moon, Kwon-Hyeon Kim, Jin Woo Lee, Jang-Joo Kim, "Influence of Host Molecules on Emitting Dipole Orientation of Phosphorescent Iridium Complexes", The International Symposium on Recent Advances and Future Issues in Organic Electroluminescence (ISOEL2016), February 17–19 (February 18), 2016, Korea (Poster).
10. Chang-Ki Moon, Kwon-Hyeon Kim, Jin Woo Lee, Jang-Joo Kim, "Influence of Host Molecules on Emitting Dipole Orientation of Phosphorescent Iridium Complexes", A-COE 2015, October 29–31 (October 30), 2015, China (Poster).
11. Chang-Ki Moon, Sei-Yong Kim, Jeong-Hwan Lee and Jang-Joo Kim, "Luminescence from oriented emitting dipoles in a birefringent thin film and OLEDs", The International Workshop on Flexible & Printable Electronics 2014 (IWEFE 2014), November 5–7, 2014, Korea. (Poster).
12. Chang-Ki Moon, Sei-Yong Kim, Jung-Bun Kim, Jeong-Hwan Lee, and Jang-Joo Kim, "Optical analysis of external quantum efficiency & total extractable light amount in OLEDs with different electrode types", KJF-ICOMEP 2013 (KJF International Conference on Organic Materials for Electronics and Photonics), August 28–31 (August 30), 2013, Korea (Poster).
13. Chang-Ki Moon, Sei-Yong Kim, Jung-Bun Kim, Jeong-Hwan Lee, and Jang-Joo Kim, "Optical analysis of external quantum efficiency & total extractable light amount in OLEDs with different electrode types", IMID 2013 (The 13th International Meeting on Information Display), August 26–29 (August 29), 2013, Korea (Poster).

List of Patents

1. Jang-Joo Kim, Chang-Ki Moon, Sei-Yong Kim, Jeong-Hwan Lee, "Apparatus for angular dependent measurement of photoluminescence emission spectrum", 10-1468065 (2014.11.26).
2. Jang-Joo Kim, Chang-Ki Moon, Sei-Yong Kim, Jeong-Hwan Lee, "Apparatus for angular dependent measurement of photoluminescence emission spectrum", PCT/KR2014/011785 (2014.12.03).
3. Jang-Joo Kim, Chang-Ki Moon, "Apparatus for testing", Application No. 16-48961 (2016.04.21).
4. Jang-Joo Kim, Chang-Ki Moon, "Apparatus for testing", PCT/KR2016/004535 (2016.04.29).

Acknowledgements

Thanks to the people who helped me, I was able to finish my degree. It was the most valuable experience of my life.

Professor Jang-Joo Kim was my advisor and an exemplary scientist. I greatly appreciate his support in my graduate school life. I will always think of his passion and conviction as a scientist.

I am grateful to my mentor, Dr. Sei-Yong Kim. His optical theory and simulation program are the basis of all my research. I would like to acknowledge Prof. Jun Yeob Lee in Sungkyunkwan University, Prof. Hironori Kaji in Kyoto University, and Dr. Jin Woo Lee from Samsung Total Petrochemicals. It was a pleasure doing wonderful collaborations and discussions with them. Dr. Andrienko kindly supported me during my three-month visit at Max Planck Institute for Polymer Research. I appreciate it, and I could start to do molecular dynamics simulations with his help. Prof. Youn Joon Jung in Seoul National University also gave me lots of advice on the molecular dynamics simulation. I especially thank Dr. Matthew D. Halls and Dr. Shaun H. Kwak from Schrodinger Inc. for their kind and helpful discussions in calculations and simulations. I could enjoy laboratory life thanks to the smart and friendly laboratory members.

Finally, I would like to thank my family who always give me love and support.

Contents

List of Figures

List of Tables

Chapter 1
Introduction

1.1 Molecular Orientation in Organic Electronics

1.1.1 History

Molecular orientation is an important factor in determining the electrical and optical properties of organic semiconductors. The molecular alignment of liquid crystals has long been significant for applications in flat panel displays [1, 2]. The alignment of polymer chains in conjugated polymer films is a well-known feature [3–5]. However, the conjugated small molecules in vacuum-deposited films has been regarded as randomly oriented for a long time because their intermolecular interactions are weak and there is no directional force in the molecular deposition. Lin et al. reported optical birefringence of vacuum-deposited oligofluorene films [6] indicating preferred alignment of the molecules in the films (Fig. 1.1). The optical birefringence was investigated from the many vacuum-deposited films composed of long or planar-shaped organic molecules [7, 8]. Intermolecular hydrogen bonds make a rigid structure and induce further molecular alignment in films [9]. Kinetics of the molecules on the film surface during the deposition were attributed to one of the reasons of the molecular alignment. Slow surface mobility restricts molecular rotation on the surface and helps the molecular alignment [10]. Lowering the substrate temperature leads to further molecular alignments by the same token [11, 12]. Figure 1.2 shows a few molecules with the preferred orientations in vacuum-deposited layers.

1.1.2 Importance of the Molecular Orientation

Orientation of molecules in molecular films dictates their electrical and optical properties such as charge mobility [13, 14], birefringence [7], absorption [15], emission [16], ionization potential [17], and dielectric [18] and ferroelectric properties [19].

C.-K. Moon, *Molecular Orientation and Emission Characteristics of Ir Complexes and Exciplex in Organic Thin Films*, Springer Theses,
https://doi.org/10.1007/978-981-13-6055-8_1

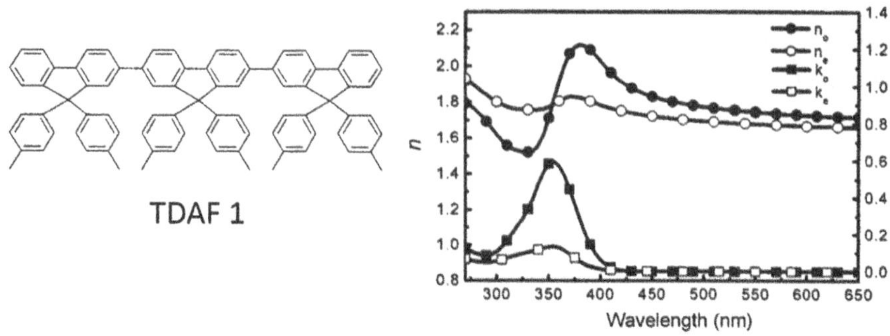

Fig. 1.1 A molecular structure of ter(9,9-diarylfluoreene) (TDAF 1) and optical constants of the TDAF 1 film

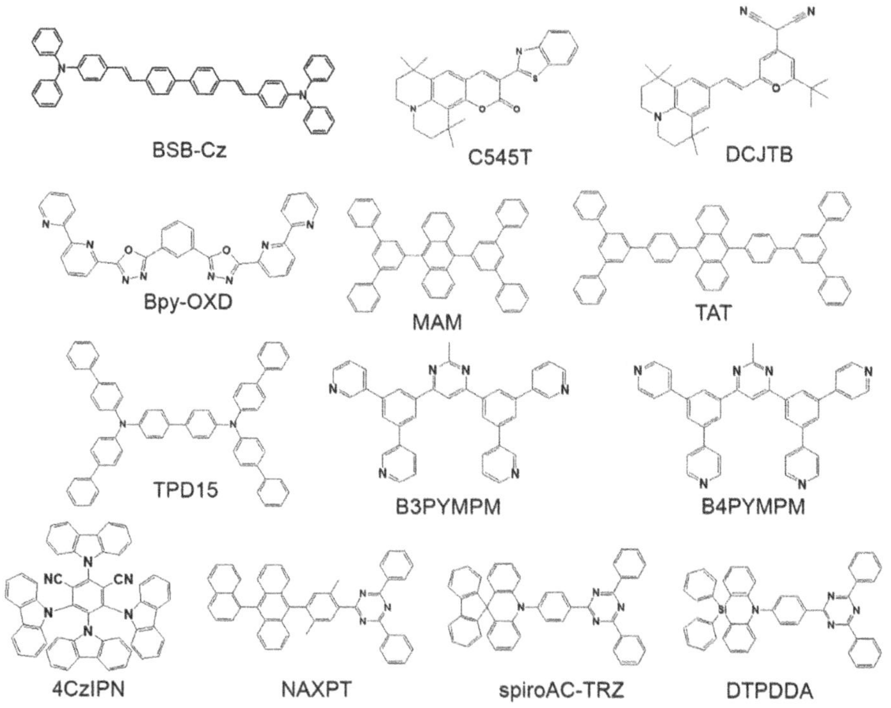

Fig. 1.2 Organic molecules have preferred orientations in vacuum-deposited films

Therefore, understanding and control of molecular orientation in organic films have been a research topic with central importance in organic electronics and photonics, including the fields of liquid crystals [1], organic field effect transistors (OTFTs) [20], and organic photovoltaics (OPVs) [21]. In organic light-emitting diodes (OLEDs), the molecular orientation of emitter embedded in the emissive layer has been an issue to enhance the outcoupling efficiency of light pursuing the horizontal alignment of the emitting dipole moment [7, 22–31].

Figure 1.3a depicts propagation of the electromagnetic wave from vertically (⊥) and horizontally (||) aligned dipoles in a OLED structure. The time-averaged Poynting vector (⟨S⟩) of the electromagnetic wave by the dipole radiation is along the direction perpendicular to the dipole moment by the equation.

$$\langle \mathrm{S} \rangle = \left(\frac{\mu_0 p_0^2 \omega^4}{32\pi^2 c} \right) \frac{\sin^2 \theta}{r^2} \hat{\mathrm{r}}, \tag{1.1}$$

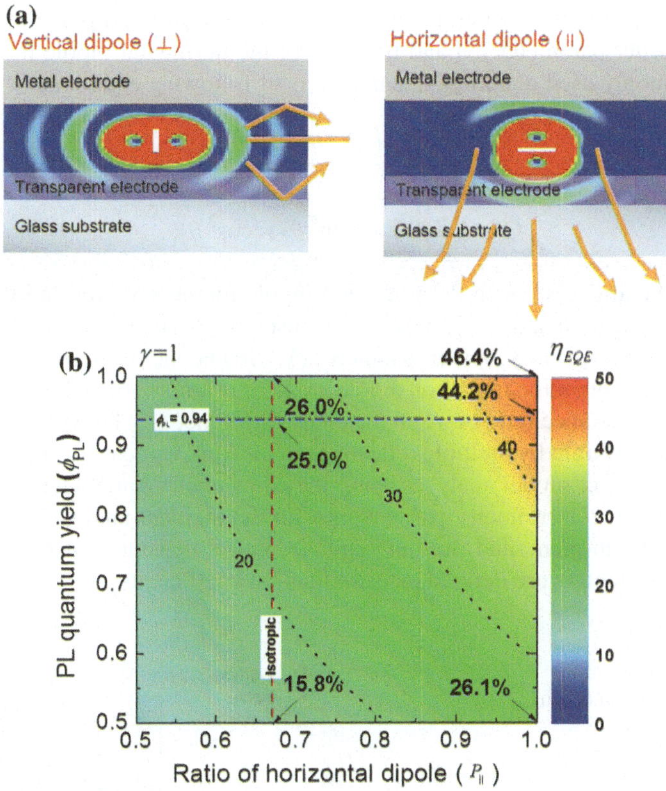

Fig. 1.3 **a** Radiation from horizontal and vertical dipoles in OLED stacks. **b** Theoretical calculation of EQE as functions of the ratio of horizontal dipole moment and radiative quantum efficiency

where μ_0 is the permeability of free space, p_0 is the dipole moment, ω is the angular frequency, c is the speed of light, θ is the angle between the dipole moment and propagating direction of the electromagnetic wave, r is the position, and $\hat{r}$ is the unit vector in the direction of r. Most of the electromagnetic waves from the vertical dipole cannot propagate to the air and they are confined to the high refractive index organic layers ($n \sim 1.8$). In contrast, the electromagnetic waves from the horizontal dipole can be outcoupled due to the low incident angle. Theoretical calculation (Fig. 1.3b) [32] proposed 46% of the outcoupling efficiency with the perfectly aligned dipole moments parallel to the substrate, whereas the isotropic dipole orientation (2/3 of horizontal and 1/3 of vertical dipole moments) only contribute to 27% of EQE.

1.1.3 Estimation of Molecular Orientations

The orientation order of the molecules in the film is represented by horizontal (in-plane) and vertical (out-of-plane) components of the transition dipole moment. The orientation order parameter (S) and the ratio of the horizontal dipole moment (Θ) are the commonly used quantitative parameters as following equations:

$$S = \frac{3\langle\cos^2\theta\rangle - 1}{2}, \tag{1.2}$$

$$\Theta : (1 - \Theta) = \langle\sin^2\theta\rangle : \langle\cos^2\theta\rangle, \tag{1.3}$$

where θ is the angle between the transition dipole moment vector and the normal vector of the substrate and $\langle\ldots\rangle$ indicates an ensemble average (Fig. 1.4).

Variable-angle spectroscopy ellipsometry (VASE) is one of the methods to estimate the orientation of the transition dipole moment in amorphous films. It measures the complex reflectance ratio (Ψ) and the phase difference (Δ) between the incident s- and p-polarized light by multiple reflections in the film. Optical constants (n and k) are determined after model analyses using the ellipsometry parameters. The model analyses using the ellipsometry parameters with the incident angles give separated solutions of the in-plane and the out-of-plane optical constants of the films. The orientation parameters are defined using extinction coefficients as

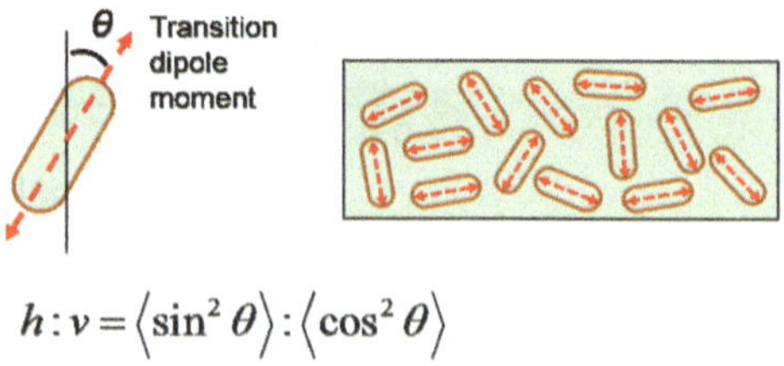

Fig. 1.4 Estimation of the molecular orientation in thin films using the transition dipole moment

$$S = \frac{k_e^{max} - k_o^{max}}{k_e^{max} + 2k_o^{max}}, \tag{1.4}$$

$$\Theta : (1 - \Theta) = 2k_o^{max} : k_e^{max}. \tag{1.5}$$

The emitting dipole orientation (EDO) describes the ratio of horizontal and vertical emitting dipole moments, whereas the extinction coefficients indicate the ratio of the absorption dipole moments. The EDO can be determined by angle-dependent PL analysis [22]. Figure 1.5a exhibits an experimental setup of the angle-dependent PL measurement. A half cylinder lens made of fused silica is fixed on a rotation stage. A film is deposited on a fused silica substrate and the substrate is attached to the lens using refractive index liquid. Angle-dependent *p*-polarized emission by an incidence of ultraviolet (UV) light with a rotation of the stage is measured at the side of the half-cylinder lens. Optical simulation of thin film luminescence determines the ratio of horizontal and vertical emitting dipole moment by the fitting of the angular intensity profile. This method can analyze the EDO of both neat emissive layers and dye-doped layers. Figure 1.5b shows an example of the analysis of bis(2-phenylpyridine)iridium-(III)(2,2,6,6-tetramethylheptane-3,5-diketonate) ($Ir(ppy)_2tmd$) doped in a 30-nm-thick N,N′-dicarbazolyl-4-4′-biphenyl (CBP) layer. The scatters indicate the angular emission intensity of $Ir(ppy)_2tmd$ and the lines are calculated by the optical simulation applying different dipole orientations. The scatters are located between the lines representing isotropic and fully horizontal orientations. The horizontal to vertical dipole ratio (*h*:*v*) of $Ir(ppy)_2tmd$ in the CBP layer is determined by 0.75:0.25.

1.2 Optical Models of OLEDs

The current optical analysis of OLEDs is based on the dipole radiation model [33]. OLEDs are composed of a stack of very thin layers comparable to the wavelength of visible light, so called microcavity. The radiative decay rate of emitting molecules in microcavity does not agree with the decay rate in free space, which is called Purcell effect [34]. The dipole radiation model describing molecular fluorescence near planar interfaces was suggested by Chance et al. [35]. This model have been developed to analyze emission of OLEDs [36–39].

An oscillating dipole near interfaces is regarded as a forced damped harmonic oscillator. The equation of the motion of the dipole was expressed as

$$m\frac{d^2x}{dt^2} + mb_0\frac{dx}{dt} + m\omega_0^2 x = eE_R \tag{1.6}$$

where m is the effective mass of the dipole, x is the displacement of the dipole, b_0 is the decay rate in free space, ω_0 is the resonant angular frequency, e is the electric

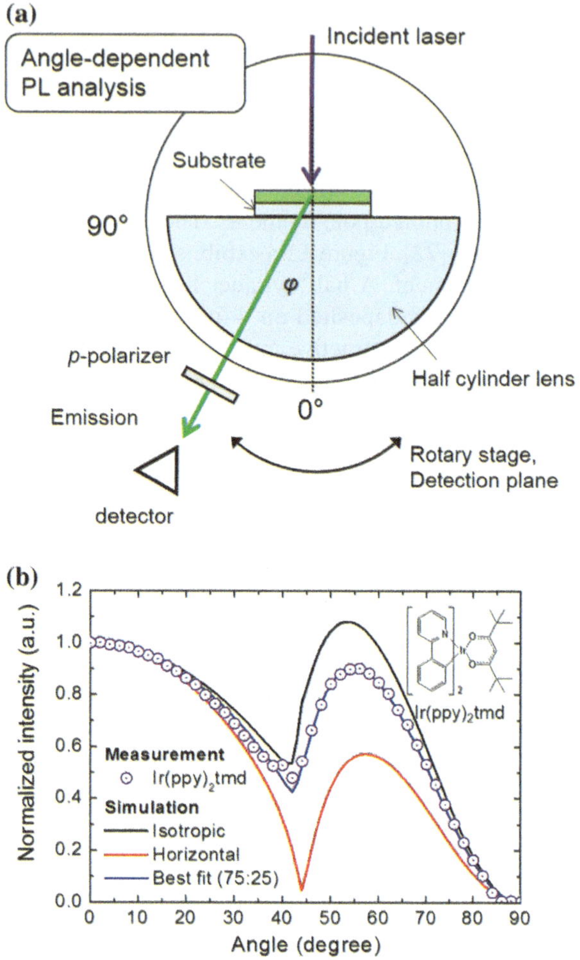

Fig. 1.5 **a** An experimental setup of angle-dependent PL measurement. **b** Angular p-polarized emission profiled of $Ir(ppy)_2tmd$ and fit by optical simulation of film luminescence

charge constant, and E_R is the reflected electric field. Otherwise, it is expressed with dipole moment ($\mu = mx$)

$$\frac{d^2\mu}{dt^2} + b_0\frac{d\mu}{dt} + \omega_0^2\mu = \frac{e^2}{m}E_R. \tag{1.7}$$

The dipole and the reflected electric field oscillate with the same complex frequency of $\Omega = (\omega_0 + \Delta\omega) - i(b/2)$, where b is the decay rate near interfaces. Then, solutions of μ and E_R are expressed as

$$\mu = \mu_0 \exp(-i\Omega t), \tag{1.8}$$

$$E_R = E_{R,0}\exp(-i\Omega t). \tag{1.9}$$

Substituting μ and E_R to Eq. (1.7) gives

$$\left(-\Omega^2 - i\Omega b_0 + \omega_0^2\right)\mu_0 = \frac{e^2}{m}E_{R,0}, \tag{1.10}$$

and the solution is

$$\begin{aligned}\Omega &= -i\frac{b_0}{2} + \omega_0\sqrt{1 - \frac{b_0^2}{4\omega_0^2} - \frac{e^2}{m\mu_0\omega_0^2}E_{R,0}} \\ &\approx -i\frac{b_0}{2} + \omega_0\left(1 - \frac{b_0^2}{8\omega_0^2} - \frac{e^2}{2m\mu_0\omega_0^2}E_{R,0}\right)\end{aligned} \tag{1.11}$$

with assumptions that $b_0^2 \ll \omega_0^2$ and $\frac{e^2}{m\mu_0\omega_0^2}E_{R,0} \ll \omega_0^2$. By dividing Ω into real and imaginary parts, b and $\Delta\omega$ are

$$b = b_0 + \frac{e^2}{m\mu_0\omega_0^2}\mathrm{Im}\left(E_{R,0}\right) \tag{1.12}$$

$$\Delta\omega = \frac{b_0^2}{8\omega_0^2} + \frac{e^2}{2m\mu_0\omega_0^2}\mathrm{Re}\left(E_{R,0}\right) \tag{1.13}$$

The clssical formula for the radiatve decay rate is [40]

$$b_r = \frac{2e^2k_1^3}{m\omega n_1^2} \tag{1.14}$$

where n_1 is the refractive index of the medium containing the dipole and k_1 is the wave vector in the medium. Because the radiative quantum efficiency is $q = b_r/b$, Eq. (1.12) can be rewritten with the normalized decay rate $\left(\hat{b}\right)$ as

$$\hat{b} = \frac{b}{b_0} = 1 + \frac{e^2}{b_0 m\mu_0\omega_0^2}\mathrm{Im}\left(E_{R,0}\right) = 1 + \frac{3qn_1^2}{2\mu_0 k_1^3}\mathrm{Im}\left(E_{R,0}\right) \tag{1.15}$$

Now, the decay rate is modified by the reflected electric field. The solution of the electric field radiated from a dipole near interfaces was investigated early by Sommerfeld [41] who used a Hertz vector potential (Π) in cylindrical coordinates. This method gives the solution of decay rate of dipole aligned vertically and horizontally to the interface. When the emitting layer is sandwiched by two layers like an OLED as illustrated in Fig. 1.6, the normalized decay rates of vertically ($\perp$) and horizontally ($\|$) aligned dipoles are given as follows:

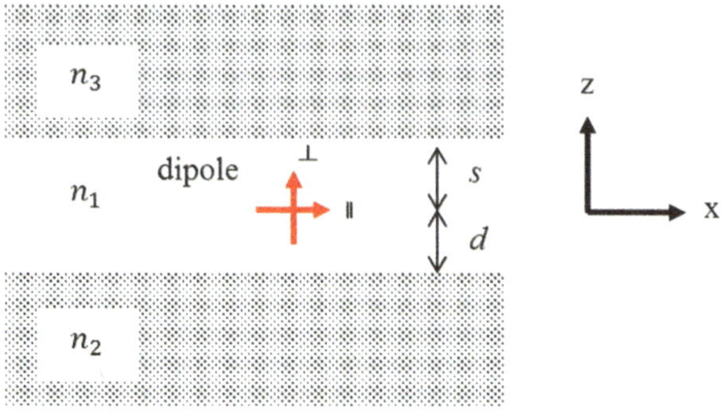

Fig. 1.6 An environment of horizontal and dipoles embedded in a alyer sandwiched by two layers

$$\hat{b}_{\perp,\parallel} = \frac{b_{\perp,\parallel}}{b_0} = 1 - qZ_{\perp,\parallel}, \tag{1.16}$$

$$Z_{\perp} = 1 - \frac{3}{2}\int_0^{\infty} \frac{\left[1 - R\left(d, r_{12}^{\mathrm{TM}}\right)\right]\left[1 - R\left(s, r_{13}^{\mathrm{TM}}\right)\right]}{1 - R\left(d + s, r_{12}^{\mathrm{TM}} r_{13}^{\mathrm{TM}}\right)} \frac{u^3}{l_1} du, \tag{1.17}$$

$$Z_{\parallel} = 1 - \frac{3}{4}\mathrm{Re}\int_0^{\infty} \left\{ \begin{array}{l} \frac{\left[1 + R\left(d, r_{12}^{\mathrm{TM}}\right)\right]\left[1 + R\left(s, r_{13}^{\mathrm{TM}}\right)\right]}{1 - R\left(d + s, r_{12}^{\mathrm{TM}} r_{13}^{\mathrm{TM}}\right)}\left(1 - u^2\right) \\ + \frac{\left[1 + R\left(d, r_{12}^{\mathrm{TE}}\right)\right]\left[1 + R\left(s, r_{13}^{\mathrm{TE}}\right)\right]}{1 - R\left(d + s, r_{12}^{\mathrm{TE}} r_{13}^{\mathrm{TE}}\right)} \end{array} \right\} \frac{u}{l_1} du, \tag{1.18}$$

where u is the normalized in-plane wave vector, $l_j = \left(n_j^2/n_1^2 - u^2\right)^{1/2}$, $R(x, y) = y\exp(2ik_1 l_1 x)$, and $r_{mn}^{\mathrm{TM/TE}}$ is the Fresnel's reflection coefficient of transverse magnetic (TM)/transverse electric (TE) waves (or p-/s-polarized light) from medium m to n. The transfer matrix method can be used to calculate the reflection coefficient of a multi-layer structure [42]. When the dipole distribution of the emitter is isotropic, the normalized decay rate is a combination of that of perpendicular and parallel dipoles, which is expressed as

$$\hat{b}_{\mathrm{iso}} = \frac{1}{3}\hat{b}_{\perp} + \frac{2}{3}\hat{b}_{\parallel} \tag{1.19}$$

The normalized decay rate also can be decribed with the relative power dissipation function (p) of the dipole.

$$\hat{b} = 1 - q + q\int_0^{\infty} p\,du \tag{1.20}$$

Therefore, the quantum efficiency of emitter in microcavity is now modified as

$$q_{\text{eff}} = \frac{b_r}{\hat{b}} = \frac{q \int_0^\infty p du}{1 - q + q \int_0^\infty p du} \tag{1.21}$$

The integration of the power disspation function corresponds to Purcell factor in the microcavity.

The light generated in OLEDs couples to several optical modes. Since organic layers in OLEDs have higher refractive indices than air, much of light is confinend inside as waveguide modes. Some of the light is reflected at the glass/air interface, some is absorbed to metal and ITO electrodes, and other escapes out to air. These optical modes can be determined by the power dissipation of the dipole according the direction of the electromanetic wave. The in-plane wave vector (u) indicates the direction of the wave. u_{air} and $u_{\text{substrate}}$, which are refractive indices of air and substrate divided by the refractive index of the emissive layer, respectively, indicate critical angles. The power dissipation function in the range of $0 \leq u \leq u_{\text{air}}$ indicates the power function of the electromagnetic wave will be radiated to air (outcoupled mode). In the same manner, if the in-plane wave vector is $u_{\text{air}} < u \leq u_{\text{substrate}}$, the wave is confined in the substrate (substrate mode), if the in-plane wave vector is $u_{\text{substrate}} < u \leq 1$, the wave is confined in the organic layer (waveguided mode), and if the in-plane wave vector is $1 < u < \infty$, the wave is coupled to suface plasmon polaritons at the organic/metal surface (SPPs mode).

1.3 Iridium Complex

Phosphorescent dyes, organometallic compounds with heavy metal atoms, have been widely used for high efficiency OLEDs. Electrical recombination in OLEDs generates the singlet and triplet exciton with 1:3 ratio by spin statistics. Since the emission from the triplet state is not allowed, only 25% of excitons participate in the emission for normal fluorescent emitters. However, phosphorescent dyes lead to emission from triplet states through spin flipping by strong spin-orbit coupling so 100% of the excitons can participate in light emission. Since Baldo et al. [43] used 2,3,7,8,12,13,17,18-octaethyl-21H,23H-porphine platinum(II) (PtOEP) as a phosphorescent dye in 1998, a variety of molecules have been developed and greatly increase the OLED efficiency. The most widely used dyes in OLEDs are iridium complexes. The iridium complex generally has three cyclometalating ligands. Homoleptic Ir complexes have equivalent ligands, and heteroleptic Ir complexes has different ligands, generally two aromatic main ligands and one aliphatic ancillary ligand. Figure 1.7 illustrates a homoleptic complex, $Ir(ppy)_3$, and two heteroleptic complexes, $Ir(ppy)_2tmd$ and $Ir(MDQ)_2acac$.

The most widely used dye in OLEDs is an iridium complex [44]. The iridium complex generally has three cyclometalating ligands. Homoleptic Ir complexes have equivalent ligands [45], and heteroleptic Ir complexes have different ligands, generally two aromatic main ligands and one aliphatic ancillary ligand [46].

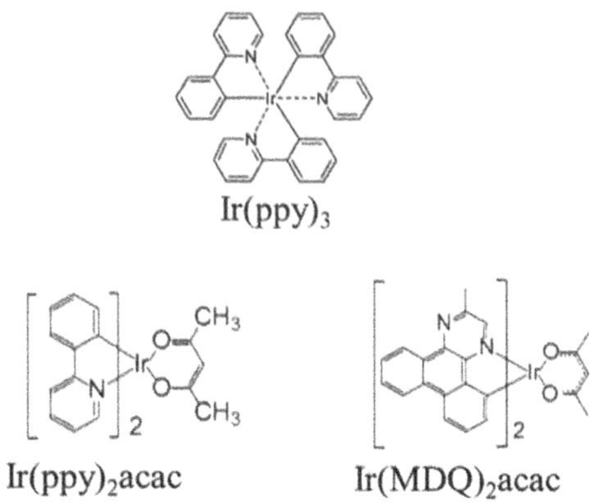

Fig. 1.7 Molecular structures of some iridium complexes

Figure 1.7 illustrates a homoleptic complex, Ir(ppy)$_3$, and two heteroleptic complexes, Ir(ppy)$_2$tmd and Ir(MDQ)$_2$acac. Ir complexes had been regarded as having isotropic dipole distribution due to their octahedral and globular molecular structures but Schmidt et al. [47] reported horizontal EDO of Ir(MDQ)$_2$acac doped in NPB layer in 2011. Since then, horizontal dipole orientations from various compounds such as Ir(ppy)$_2$acac [24, 48], Ir(bppo)$_2$acac [30], Ir(ppy)$_2$tmd [49, 50], Ir(mphmq)$_2$tmd [26], FIrpic [51], tBu-FCNIr [52], Ir(piq)$_3$ [30, 33] and etc. were investigated by many groups.

1.4 Exciplex

Exciplex is an acronym of excited state charge transfer complex. Relaxation of am exciplex emits a photon that has lower energy in compared to the photon energy from monomer emission. The redshift emission spectrum is usually an indicator of the exciplex. Figure 1.8a exhibits formation of an exciplex by an intermolecular charge transfer between the electron donor and acceptor. Figure 1.8b schematically depicts an exciplex emission spectrum compared to the absorption and emission spectra of the monomer.

Recently, the application of exicplex has attracted attention in OLEDs. Frontier orbitals of an exciplex is separated into the HOMO of donor and the LUMO of acceptor, respectively. The low overlap of the frontier orbitals leads to very low energy gap between singlet and triplet excited states because of low electron repulsion by the electron exchange. As a result, intersystem crossing and reverse intersystem crossing between the singlet and triplet states are active at room temperature (Fig. 1.9a). Therefore, fast prompt fluorescence ($\tau_p \sim 100$ ns) and delayed fluorescence ($\tau_d \sim 100$ μs) are observed in the transient PL decay curve (Fig. 1.9b). The spin mixing between singlet and triplet states enables exploiting triplet excited states in OLEDs under

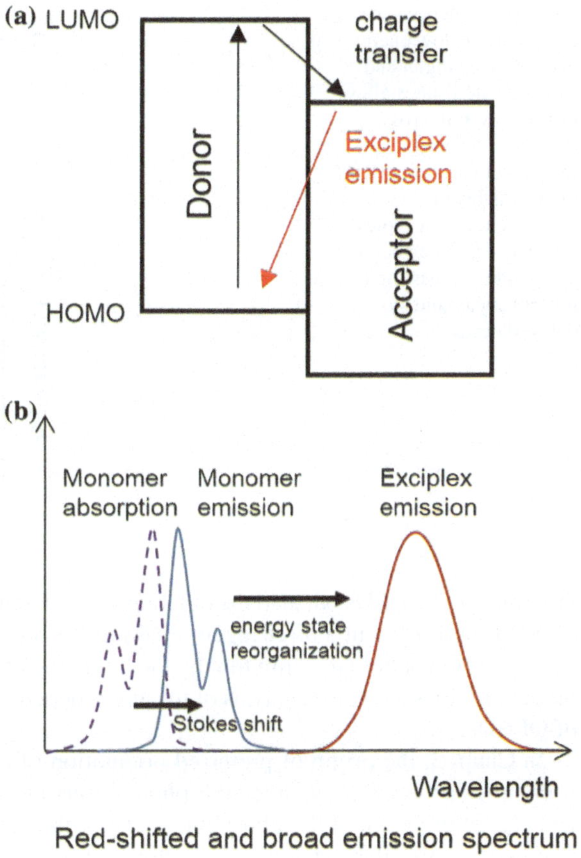

Fig. 1.8 **a** Charge transfer between donor and acceptor upon excitation to form an exciplex. **b** The emission spectrum of exciplex which is red-shifted in compared to absorption and emission of monomers

electrical operation as delayed fluorescence [53], leading highly efficient OLEDs comparable to phosphorescent OLEDs [54].

1.5 Outline of the Thesis

This thesis describes molecular orientations of the Ir complexes and the exciplex in organic thin films with an optical model of luminescence of from an anisotropic microcavity.

In Chap. 2, an optical model to describe the luminescence from oriented emitting dipoles in a birefringent medium is presented. The theoretical model is validated by applications to a dye doped organic thin film and OLEDs. This thesis demonstrates that the optical birefringence affects not only far-field radiation characteristics such as the angle-dependent photoluminescence and electroluminescence spectra from

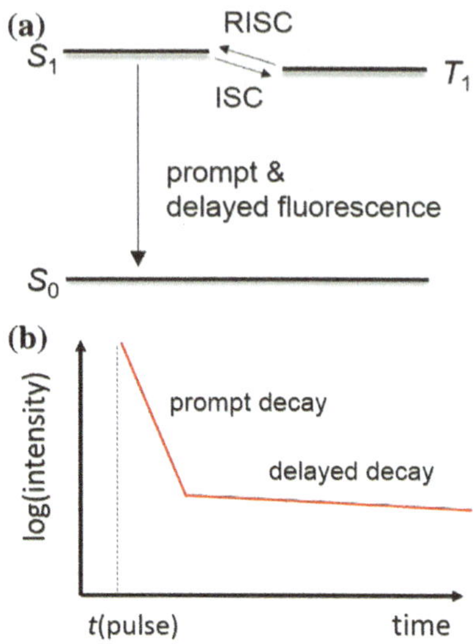

Fig. 1.9 **a** Energy state of exciplex. The low energy gap between the singlet and triplet excited states allow fast intersystem crossing (ISC) and reverse intersystem crossing (RISC), thereby delayed fluorescence. **b** A typical transient PL decay curve of an exciplex consisting of fast prompt decay and slow delayed decay

thin films and OLEDs but also the outcoupling efficiency of OLEDs. The orientation of emitting dipoles in a birefringent medium is successfully analyzed from the far-field radiation pattern of a thin film by the birefringent model. In addition, the optical model provides a precise analysis of the angle-dependent EL spectra and efficiencies of OLEDs.

In Chap. 3, the origin of preferred orientation of emitting dipole of heteroleptic Ir complexes doped in organic amorphous layers are revealed by quantum mechanical characterization of the phosphors along with the simulation of vacuum deposition using molecular dynamics for a direct comparison with experimental observations of EDO. Consideration of both the electronic transitions in a molecular frame and the orientation of the molecules interacting with the environment at the vacuum/molecular film interface allows quantitative analyses of the EDO depending on host molecules and dopant structures. The interaction between the phosphor and nearest host molecules on the surface, minimizing the non-bonded interaction energy determines the molecular alignment during the vacuum deposition. Parallel alignment of the main cyclometalating ligands in the molecular complex due to host interactions leads to the horizontal EDO rather than the ancillary ligand orienting to vacuum.

In Chap. 4, the electronic structure and emission process of exciplex is presented considering the donor-acceptor heterodimer configurations in solid states. PL properties of the exciplex depend on the exciplex energy. Quantum mechanical characterization reveals that the various dimer configurations in the solid lead to inhomogeneous broadening of the emission spectrum and excited state dynamics. The spectral redshift of the exciplex in transient PL as time delays are interpreted by superposition of the fast-decaying high-energy exciplex and slow-decaying low-energy exciplex in the blend. An energy diagram of the exicplex with band-like singlet and triplet states is presented based on the experimental and calculation results.

References

1. Raynes P (1993) LIQUID CRYSTALS—Second Edition, by S CHANDRASEKHAR, Cambridge University Press, (1992), ISBN 0-521-41747-3 (HB), ISBN 0-521-42741-X (PB). Liq Cryst Today 3:7–7. https://doi.org/10.1080/13583149308628627
2. Yeh P, Gu C (2010) Optics of liquid crystal displays, vol 67. Wiley
3. McBranch D, Campbell I, Smith D, Ferraris J (1995) Optical determination of chain orientation in electroluminescent polymer films. Appl Phys Lett 66:1175–1177
4. Kim J-S, Ho PKH, Greenham NC, Friend RH (2000) Electroluminescence emission pattern of organic light-emitting diodes: implications for device efficiency calculations. J Appl Phys 88:1073–1081. https://doi.org/10.1063/1.373779
5. Smith LH, Wasey JAE, Samuel IDW, Barnes WL (2005) Light out-coupling efficiencies of organic light-emitting diode structures and the effect of photoluminescence quantum yield. Adv Func Mater 15:1839–1844. https://doi.org/10.1002/adfm.200500283
6. Lin H-W et al (2004) Anisotropic optical properties and molecular orientation in vacuum-deposited ter(9,9-diarylfluorene)s thin films using spectroscopic ellipsometry. J Appl Phys 95:881–886. https://doi.org/10.1063/1.1635991
7. Yokoyama D (2011) Molecular orientation in small-molecule organic light-emitting diodes. J Mater Chem 21:19187. https://doi.org/10.1039/c1jm13417e
8. Yokoyama D, Sakaguchi A, Suzuki M, Adachi C (2008) Horizontal molecular orientation in vacuum-deposited organic amorphous films of hole and electron transport materials. Appl Phys Lett 93:173302. https://doi.org/10.1063/1.2996258
9. Sasabe H et al (2011) Influence of substituted pyridine rings on physical properties and electron mobilities of 2-methylpyrimidine skeleton-based electron transporters. Adv Func Mater 21:336–342
10. Mayr C, Brütting W (2015) Control of molecular dye orientation in organic luminescent films by the glass transition temperature of the host material. Chem Mater 27:2759–2762. https://doi.org/10.1021/acs.chemmater.5b00062
11. Komino T, Tanaka H, Adachi C (2014) Selectively controlled orientational order in linear-shaped thermally activated delayed fluorescent dopants. Chem Mater 26:3665–3671. https://doi.org/10.1021/cm500802p
12. Komino T, Nomura H, Koyanagi T, Adachi C (2013) Suppression of efficiency roll-off characteristics in thermally activated delayed fluorescence based organic light-emitting diodes using randomly oriented host molecules. Chem Mater 25:3038–3047. https://doi.org/10.1021/cm4011597
13. Sirringhaus H et al (1999) Two-dimensional charge transport in self-organized, high-mobility conjugated polymers. Nature 401:685–688
14. Sundar VC et al (2004) Elastomeric transistor stamps: reversible probing of charge transport in organic crystals. Science 303:1644–1646
15. Erb T et al (2005) Correlation between structural and optical properties of composite polymer/fullerene films for organic solar cells. Adv Func Mater 15:1193–1196
16. Taminiau T, Stefani F, Segerink F, Van Hulst N (2008) Optical antennas direct single-molecule emission. Nat Photonics 2:234–237
17. Duhm S et al (2008) Orientation-dependent ionization energies and interface dipoles in ordered molecular assemblies. Nat Mater 7:326–332
18. Yamao T, Yamamoto K, Taniguchi Y, Miki T, Hotta S. (2008) Laser oscillation in a highly anisotropic organic crystal with a refractive index of 4.0. J Appl Phys 103:093115
19. Horiuchi S, Tokura Y (2008) Organic ferroelectrics. Nat Mater 7:357–366

20. Horowitz G (1998) Organic field-effect transistors. Adv Mater 10:365–377
21. Janssen RA, Nelson J (2013) Factors limiting device efficiency in organic photovoltaics. Adv Mater 25:1847–1858
22. Frischeisen J, Yokoyama D, Adachi C, Brütting W (2010) Determination of molecular dipole orientation in doped fluorescent organic thin films by photoluminescence measurements. Appl Phys Lett 96:073302. https://doi.org/10.1063/1.3309705
23. Flämmich M et al (2011) Oriented phosphorescent emitters boost OLED efficiency. Org Electron 12:1663–1668
24. Liehm P et al (2012) Comparing the emissive dipole orientation of two similar phosphorescent green emitter molecules in highly efficient organic light-emitting diodes. Appl Phys Lett 101:253304
25. Kim S-Y et al (2013) Organic light-emitting diodes with 30% external quantum efficiency based on a horizontally oriented emitter. Adv Func Mater 23:3896–3900. https://doi.org/10.1002/adfm.201300104
26. Kim KH et al (2014) Phosphorescent dye-based supramolecules for high-efficiency organic light-emitting diodes. Nat Commun 5:4769. https://doi.org/10.1038/ncomms5769
27. Mayr C et al (2014) Efficiency enhancement of organic light-emitting diodes incorporating a highly oriented thermally activated delayed fluorescence emitter. Adv Func Mater 24:5232–5239. https://doi.org/10.1002/adfm.201400495
28. Kim KH et al (2015) Controlling emitting dipole orientation with methyl substituents on main ligand of iridium complexes for highly efficient phosphorescent organic light-emitting diodes. Adv Opt Mater 3:1191–1196
29. Kim K-H, Ahn ES, Huh J-S, Kim Y-H, Kim J-J (2016) Design of heteroleptic Ir complexes with horizontal emitting dipoles for highly efficient organic light-emitting diodes with an external quantum efficiency of 38%. Chem Mater 28:7505–7510
30. Jurow MJ et al (2016) Understanding and predicting the orientation of heteroleptic phosphors in organic light-emitting materials. Nat Mater 15:85–91. https://doi.org/10.1038/nmat4428
31. Komino T et al (2016) Electroluminescence from completely horizontally oriented dye molecules. Appl Phys Lett 108:241106. https://doi.org/10.1063/1.4954163
32. Kim SY et al (2013) Organic light-emitting diodes with 30% external quantum efficiency based on a horizontally oriented emitter. Adv Func Mater 23:3896–3900
33. Moon C-K (2014) Analysis of orientation of the transition dipole moment of phosphorescent dyes in the energy-transferring condition. Master thesis, Seoul National University
34. Purcell EM (1946) Spontaneous emission probabilities at radio frequnecy. Phys Rev 69:681
35. Chance RR, Prock A, Silbey R (1978) Molecular fluorescence and energy transfer near interfaces. Adv Chem Phys 37:1–65
36. Barnes WL (1998) Fluorescence near interfaces: the role of photonic mode density. J Mod Opt 45:661–699. https://doi.org/10.1080/09500349808230614
37. Wasey JAE, Barnes WL (2000) Efficiency of spontaneous emission from planar microcavities. J Mod Opt 47:725–741. https://doi.org/10.1080/09500340008233393
38. Neyts K (2005) Microcavity effects and the outcoupling of light in displays and lighting applications based on thin emitting films. Appl Surf Sci 244:517–523. https://doi.org/10.1016/j.apsusc.2004.09.156
39. Furno M, Meerheim R, Hofmann S, Lüssem B, Leo K (2012) Efficiency and rate of spontaneous emission in organic electroluminescent devices. Phys Rev B 85. https://doi.org/10.1103/physrevb.85.115205
40. Kuhn H (1970) Classical aspects of energy transfer in molecular systems. J Chem Phys 53:101–108. https://doi.org/10.1063/1.1673749
41. Tang CW, VanSlyke SA (1987) Organic electroluminescent diodes. Appl Phys Lett 51:913. https://doi.org/10.1063/1.98799
42. Charalambos C, Katsidis DIS (2002) General transfer-matrix method for optical multilayer systems with coherent, partially coherent, and incoherent interference. Appl Opt 41:4978
43. Baldo MA et al (1998) Highly efficient phosphorescent emission from organic electroluminescent devices. Nature 395:151–154

44. Baldo MA, Lamansky S, Burrows PE, Thompson ME, Forrest SR (1999) Very high-efficiency green organic light-emitting devices based on electrophosphorescence. Appl Phys Lett 75:4–6. https://doi.org/10.1063/1.124258
45. Tsuboyama A et al (2003) Homoleptic cyclometalated iridium complexes with highly efficient red phosphorescence and application to organic light-emitting diode. J Am Chem Soc 125:12971–12979
46. Lamansky S et al (2001) Highly phosphorescent bis-cyclometalated iridium complexes: synthesis, photophysical characterization, and use in organic light emitting diodes. J Am Chem Soc 123:4304–4312
47. Schmidt TD et al (2011) Evidence for non-isotropic emitter orientation in a red phosphorescent organic light-emitting diode and its implications for determining the emitter's radiative quantum efficiency. Appl Phys Lett 99:163302. https://doi.org/10.1063/1.3653475
48. Lampe T et al (2016) Dependence of phosphorescent emitter orientation on deposition technique in doped organic films. Chem Mater 28:712–715. https://doi.org/10.1021/acs.chemmater.5b04607
49. Kim KH, Moon CK, Lee JH, Kim SY, Kim JJ (2014) Highly efficient organic light-emitting diodes with phosphorescent emitters having high quantum yield and horizontal orientation of transition dipole moments. Adv Mater 26:3844–3847. https://doi.org/10.1002/adma.201305733
50. Moon C-K, Kim K-H, Lee JW, Kim J-J (2015) Influence of host molecules on emitting dipole orientation of phosphorescent iridium complexes. Chem Mater 27:2767–2769. https://doi.org/10.1021/acs.chemmater.5b00469
51. Shin H et al (2014) Blue phosphorescent organic light-emitting diodes using an exciplex forming co-host with the external quantum efficiency of theoretical limit. Adv Mater 26:4730–4734
52. Lee JH et al (2015) Finely tuned blue iridium complexes with varying horizontal emission dipole ratios and quantum yields for phosphorescent organic light-emitting diodes. Adv Opt Mater 3:211–220
53. Goushi K, Adachi C (2012) Efficient organic light-emitting diodes through up-conversion from triplet to singlet excited states of exciplexes. Appl Phys Lett 101:023306. https://doi.org/10.1063/1.4737006
54. Liu W et al (2016) Novel strategy to develop exciplex emitters for high-performance OLEDs by employing thermally activated delayed fluorescence materials. Adv Func Mater 26:2002–2008. https://doi.org/10.1002/adfm.201505014

Chapter 2
Modeling of the Dipole Radiation in an Anisotropic Microcavity

2.1 Introduction

Light emission from organic materials has been an important research topic during the last few decades because of its scientific and technological importance, particularly due to the success of organic light emitting diodes (OLEDs). In many studies, the classical dipole model is used to understand emission characteristics from thin films and devices [1–6]. This model has been employed in the prediction of emission patterns [1, 7–11], mode analysis including the outcoupling efficiency from thin films and devices [1, 7, 8, 12–22], position of recombination zones in electroluminescent devices [14], and investigating the orientation of emitting dipoles in thin films [12, 23–28]. Until recently, the emitting dipoles were commonly assumed to be randomly oriented in an isotropic medium for small molecule based organic EL devices because there are no apparent driving forces that would yield a preferred dipole orientation for the emitters. Recently, however, a large number of organic thin films have been reported to possess a preferred orientation, leading to optical birefringence [29–32]. Furthermore, some emitters doped in organic semiconducting layers have demonstrated a preferentially oriented transition dipole moment along the horizontal direction (parallel to the substrate) [12, 23–28]. For these systems, the assumption of a random orientation of the dipoles in an isotropic medium no longer applies and thus leads to inaccurate predictions for the emission patterns (intensity and spectrum) and mode analysis in a thin film and related devices. Wasey et al. [5]. attempted to analyze emission in a polymer layer while accounting for birefringence. They proved that emitting dipoles are aligned horizontally in a spin coated thin polymer film by the analysis of far-field radiation from the film. Further research analyzed dipole radiation in the anisotropic medium with an arbitrary optical axis [33]. However, unfortunately, they did not perform the quantitative analysis of the dipole orientation in a birefringent medium and did not correlate the dipole orientation and the optical birefringence with the outcoupling efficiency of OLEDs.

C.-K. Moon, *Molecular Orientation and Emission Characteristics of Ir Complexes and Exciplex in Organic Thin Films*, Springer Theses,
https://doi.org/10.1007/978-981-13-6055-8_2

In this thesis, we present an optical model originally developed by Chance el al. [2]. To describe the luminescence from emitting dipoles in a birefringent medium and validate the theoretical model through its applications to a dye doped organic thin films and OLEDs to describe the far-field radiation, outcoupling efficiency, and orientation of emitting dipoles.

2.2 Theoretical Background

2.2.1 *Dipole Radiation in a Birefringent Medium*

Consider a dipole embedded in an infinite anisotropic medium that has a refractive index tensor. The medium has an optical axis parallel to the z-axis with an ordinary index and extraordinary index such that

$$\tilde{\tilde{n}} = \begin{pmatrix} n_x & 0 & 0 \\ 0 & n_x & 0 \\ 0 & 0 & n_z \end{pmatrix}.$$

When the dipole is embedded in this anisotropic medium, the radiated power from the dipole in the absence of an interface is given by [2].

$$P_0^{\perp} = Const \cdot n_z, \tag{2.1}$$

$$P_0^{\parallel} = Const \cdot n_x \frac{3n_x^2 + n_z^2}{4n_x^2}, \tag{2.2}$$

where $Const = \mu_0^2 \omega^4 / 12\pi \varepsilon_0 c^3$, μ_0 is the dipole moment, ω is the oscillation frequency, c is the speed of light, and $\perp$ and $\parallel$ represent a vertical and horizontal orientation of the dipole, respectively. If the ordinary refractive index is larger than the extraordinary refractive index (negative birefringence), the radiated power from the horizontal dipole is larger than that of the vertical dipole in the birefringent emitting layer, and vice versa. The net radiated power from the dipole in the absence of an interface is described by taking the dipole distribution as

$$P_0 = (1 - \alpha) \cdot P_0^{\perp} + \alpha \cdot P_0^{\parallel}, \tag{2.3}$$

where α is the ratio of the horizontal dipole (2/3 for an isotropic orientation).

If the emitting layer is sandwiched by two layers, as shown in Fig. 2.1, the spontaneous decay rate of the dipole and the radiation power are modified according to the Purcell effect [34] as follows,

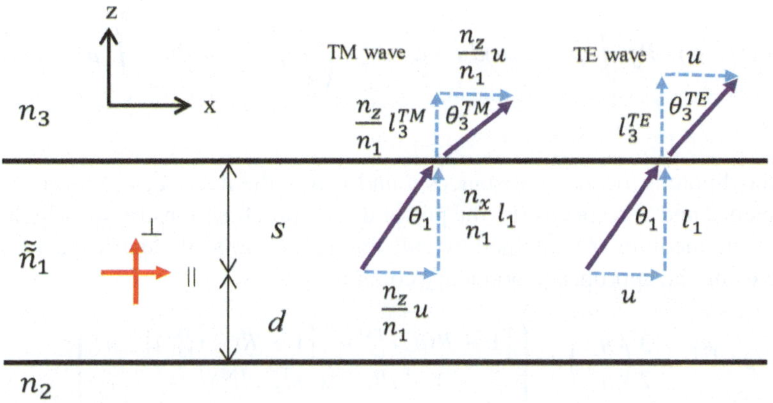

Fig. 2.1 Dipole embedded in an anisotropic medium sandwiched by two layers. The y-axis is perpendicular to the x-z plane. Arrows represent the relation between wave vector components for different electric field polarizations

$$F = \frac{b_r}{b_{r,0}} = \frac{P}{P_0}, \tag{2.4}$$

In the above expression, F is the Purcell factor, while b_r ($b_{r,0}$) and P (P_0) are the radiative decay rate and the radiated power from the dipole in the structure (in free space), respectively. In the same manner, the Purcell factor for the outcoupled power, F_{out}, can be defined as the ratio of the outcoupled power, P_{out}, to the radiated power in free space, P_0, as

$$F_{out} = \frac{P_{out}}{P_0}. \tag{2.5}$$

Modification of the radiated power from the dipole in the structure can be described by the integration of the power dissipation function $p(u)$: [2].

$$\frac{P^{\perp,\parallel}}{P_0} = 1 + \frac{2n^2}{2\mu_0 k^3}\mathrm{Im}(E_R^{\perp,\parallel}) = \int_0^\infty p^{\perp,\parallel}(u)du, \tag{2.6}$$

where k is the wavenumber, u is the normalized in-plane wave vector, and E_R denotes the reflected electric field at the position of the dipole. The net radiated power emitted by the dipole in the structure is described by the power dissipation function for each dipole orientation and electric field polarization as

$$P = (1-\alpha)\cdot P^{\perp} + \alpha \cdot P^{\parallel}$$

$$= (1-\alpha)\cdot P_0^{\perp}\int_0^{\infty} p^{\perp,\mathrm{TM}}(u)du + \alpha\cdot P_0^{\parallel}\cdot\left(\int_0^{\infty} p^{\parallel,\mathrm{TM}}(u)du + \int_0^{\infty} p^{\parallel,\mathrm{TE}}(u)du\right). \tag{2.7}$$

If the dipole is located at distances d and s from the interfaces of layers 1 and 2, and layers 2 and 3, respectively, the power dissipation functions for the dipole in the anisotropic medium are obtained by calculating the radiated electric field from the dipole using the appropriate boundary conditions as

$$p^{\perp,\mathrm{TM}} = \frac{3}{2}\left(\frac{n_z}{n_x}\right)\mathrm{Re}\left\{\frac{\left[1-R(d,r_{12}^{\mathrm{TM}})\right]\cdot\left[1-R(s,r_{13}^{\mathrm{TM}})\right]}{1-R(d+s,r_{12}^{\mathrm{TM}}r_{13}^{\mathrm{TM}})}\cdot\frac{u^3}{l_1}\right\}, \tag{2.8}$$

$$p^{\parallel,\mathrm{TM}} = \left(\frac{3n_x^2}{3n_x^2+n_z^2}\right)\mathrm{Re}\left\{\begin{array}{c}\left(\frac{n_z^2}{n_x^2}\right)\cdot\frac{\left[1+R(d,r_{12}^{\mathrm{TM}})\right]\cdot\left[1+R(s,r_{13}^{\mathrm{TM}})\right]}{1-R(d+s,r_{12}^{\mathrm{TM}}r_{13}^{\mathrm{TM}})}\\ \cdot(1-u^2)\cdot\frac{u}{l_1}\end{array}\right\}, \tag{2.9}$$

$$p^{\parallel,\mathrm{TE}} = \left(\frac{3n_x^2}{3n_x^2+n_z^2}\right)\mathrm{Re}\left\{\frac{\left[1+R(d,r_{12}^{\mathrm{TE}})\right]\cdot\left[1+R(s,r_{13}^{\mathrm{TE}})\right]}{1-R(d+s,r_{12}^{\mathrm{TE}}r_{13}^{\mathrm{TE}})}\cdot\frac{u}{l_1}\right\}, \tag{2.10}$$

where $l_j = (n_j^2/n_1^2 - u^2)^{1/2}$, $r^{\mathrm{TM,TE}}$ are the Fresnel's reflection coefficients of the TM and TE wave at the interfaces as described in Method section, and R denotes a reflection coefficient that includes a phase shift in the anisotropic medium containing the dipole described by $R(x,y) = y\cdot\exp(2ik_0 l_1 n_x x)$. Wave propagation in the anisotropic medium is affected by the refractive index of the medium parallel to the direction of electric field polarization. The relation between the wave vector components and the propagation angle is illustrated in Fig. 2.1. If the out-of-plane wave vector l_1 is real, the wave vector components of a TE wave obey the relations

$$\sin\theta_1 = u, \tag{2.11}$$

$$\cos\theta_1 = l_1, \tag{2.12}$$

while the relations for a TM wave are

$$\sin\theta_1 = \frac{n_z}{n_1}u, \tag{2.13}$$

$$\cos\theta_1 = \frac{n_x}{n_1}l_1. \tag{2.14}$$

Here, n_1 is the effective refractive index of layer 1 calculated using the refractive index ellipsoid when the power propagation angle measured from the substrate is θ_1.

To calculate the outcoupled power, p_{out}, we decompose the power dissipation function into contributions from the positive and negative z-directions for a real out-of-plane wave vector l_1 as follows [4]:

$$\begin{aligned}
&\mathrm{Re}\left(\frac{\{1 \pm R(d, r_{12})\} \cdot \{1 \pm R(s, r_{13})\}}{1 - R(d + s, r_{12}r_{13})}\right) \\
&= \frac{1}{2}\left|\frac{1 \pm R(d, r_{12})}{1 - R(d + s, r_{12}r_{13})}\right|^2 \cdot (1 - |R(s, r_{13})|^2) \\
&+ \frac{1}{2}\left|\frac{1 \pm R(d, r_{13})}{1 - R(d + s, r_{12}r_{13})}\right|^2 \\
&\cdot (1 - |R(s, r_{12})|^2).
\end{aligned} \tag{2.15}$$

The first and the second terms on the right-hand side of the above expression represent the power propagating in the positive and negative z-direction, respectively. Using the relation between reflectance, transmittance, and absorption, $1 - |R|^2 = T + A$, we divide the outcoupled power of the dipole to layer 3 using Eqs. (2.8)–(2.10) as

$$p_{\text{out}}^{\perp,\text{TM}} = \frac{3}{4}\left(\frac{n_z}{n_x}\right)\left|\frac{1 - R(d, r_{12}^{\text{TM}})}{1 - R(d + s, r_{12}^{\text{TM}} r_{13}^{\text{TM}})}\right|^2 \cdot T_{13}^{\text{TM}} \cdot \mathrm{Re}\left(\frac{u^3}{l_1}\right), \tag{2.16}$$

$$\begin{aligned}
p_{\text{out}}^{\parallel,\text{TM}} &= \frac{1}{2}\left(\frac{3n_x^2}{3n_x^2 + n_z^2}\right)\left(\frac{n_z^2}{n_x^2}\right)\left|\frac{1 + R(d, r_{12}^{\text{TM}})}{1 - R(d + s, r_{12}^{\text{TM}} r_{13}^{\text{TM}})}\right|^2 \\
&\cdot T_{13}^{\text{TM}} \cdot \mathrm{Re}\left[(1 - u^2) \cdot \frac{u}{l_1}\right],
\end{aligned} \tag{2.17}$$

$$p_{\text{out}}^{\parallel,\text{TE}} = \frac{1}{2}\left(\frac{3n_x^2}{3n_x^2 + n_z^2}\right)\left|\frac{1 + R(d, r_{12}^{\text{TE}})}{1 - R(d + s, r_{12}^{\text{TE}} r_{13}^{\text{TE}})}\right|^2 \cdot T_{13}^{\text{TE}} \cdot \mathrm{Re}\left(\frac{u}{l_1}\right), \tag{2.18}$$

where the transmittance of the TM and TE waves ($T_{13}^{\text{TM,TE}}$) in this case are described in Appendix. The outcoupled power is calculated by integrating the outcoupled power up to the critical angles. Since the critical angles for an external medium consisting of air are $u_{\text{air}}^{\text{TM}} = 1/n_z$ and $u_{\text{air}}^{\text{TE}} = 1/n_x$ for the birefringent medium, P_{out} is described by

$$\begin{aligned}
P_{\text{out}} &= (1 - \alpha) \cdot P_{\text{out}}^{\perp} + \alpha \cdot P_{\text{out}}^{\parallel} \\
&= (1 - \alpha) \cdot P_0^{\perp} \int_0^{u_{\text{air}}^{\text{TM}}} p_{\text{out}}^{\perp,\text{TM}}(u)du + \alpha \cdot P_0^{\parallel} \\
&\cdot \left[\int_0^{u_{\text{air}}^{\text{TM}}} p_{\text{out}}^{\parallel,\text{TM}}(u)du + \int_0^{u_{\text{air}}^{\text{TE}}} p_{\text{out}}^{\parallel,\text{TE}}(u)du\right].
\end{aligned} \tag{2.19}$$

2.2.2 *Efficiency of OLEDs with a Birefringent Emissive Layer*

The EQE of OLEDs is defined as the quantum ratio of the number of the photons emitted from the structure to air to the number of injected charge carriers. The emission process in the device can be divided into 4 steps and the EQE can be described as an integration of those elements [1]:

$$\eta_{\mathrm{EQE}} = \gamma \cdot \chi \cdot \int_{\lambda} s(\lambda) \cdot q_{\mathrm{eff}}(\lambda) \cdot \eta_{\mathrm{out}}(\lambda) d\lambda. \tag{2.20}$$

Here γ is the electrical balance factor (# of generated excitons/# of injected charge carriers), χ is the ratio of radiative excitons with spin statics (# of radiative excitons/# of generated excitons), $s(\lambda)$ is the normalized photon spectrum of the emitter (# of radiative excitons with λ/# of radiative excitons) satisfying $\int_{\lambda} s(\lambda) d\lambda = 1$, $q_{\mathrm{eff}}(\lambda)$ is the effective radiative quantum efficiency (# of emitted photons with λ/# of radiative excitons with E_{λ}), and $\eta_{\mathrm{out}}(\lambda)$ is the outcoupling efficiency (# of emitted photons to air/# of emitted photons with wavelength λ). Electroluminescence in the organic devices can be attributed to spontaneous emission from excitons generated through electrical operation.

Because of the modification of the radiative decay rate in the structure, the effective radiative quantum efficiency and the outcoupling efficiency of an organic EL device are described in terms of the Purcell factors as:

$$q_{\mathrm{eff}}(\lambda) = \frac{b_{\mathrm{r}}(\lambda)}{b_{\mathrm{nr}} + b_{\mathrm{r}}(\lambda)} = \frac{q \cdot F(\lambda)}{1 - q + q \cdot F(\lambda)}, \tag{2.21}$$

$$\eta_{\mathrm{out}}(\lambda) = \frac{P_{\mathrm{out}}(\lambda)}{P(\lambda)} = \frac{F_{\mathrm{out}}(\lambda)}{F(\lambda)}. \tag{2.22}$$

The EQE is then obtained using

$$\eta_{\mathrm{EQE}} = \gamma \cdot \chi \cdot \int_{\lambda} s(\lambda) \cdot \frac{q \cdot F_{\mathrm{out}}(\lambda)}{1 - q + q \cdot F(\lambda)} d\lambda. \tag{2.23}$$

The Purcell factors are be obtained by calculating the radiated power emitted by the dipole in the absence of an interface P_0, in the structure P, and the outcoupled power P_{out} as given by Eqs. (2.3), (2.7), and (2.19), respectively.

2.2.3 Far-Field Radiation

In the expression for the outcoupled power, the in-plane wave vector u can be converted to a solid angle. Then, Eq. (2.19) becomes a function of the solid angle in layer 3 (θ_3) as

$$P_{\mathrm{out}} = \int_0^{\pi/2} \left[\begin{array}{c} (1-\alpha)\cdot P_0^{\perp}\cdot p_{\mathrm{out}}^{\prime\perp,\mathrm{TM}}(\theta_3) + \alpha\cdot P_0^{\parallel} \\ \cdot\left\{ p_{\mathrm{out}}^{\prime\parallel,\mathrm{TM}}(\theta_3) + p_{\mathrm{out}}^{\prime\parallel,\mathrm{TE}}(\theta_3) \right\} \end{array} \right] \sin\theta_3 d\theta_3, \quad (2.24)$$

and the outcoupled powers as a function of θ_3 are given by

$$p_{\mathrm{out}}^{\prime\mathrm{TM}}(\theta_3) = p_{\mathrm{out}}^{\mathrm{TM}}(u)\cdot\frac{n_3 l_3^{\mathrm{TM}}}{n_z u}, \quad (2.25)$$

$$p_{\mathrm{out}}^{\prime\mathrm{TE}}(\theta_3) = p_{\mathrm{out}}^{\mathrm{TE}}(u)\cdot\frac{n_3 l_3^{\mathrm{TE}}}{n_x u}, \quad (2.26)$$

where $l_3^{\mathrm{TM}} = (n_3^2/n_z^2 - u^2)^{1/2}$ and $l_3^{\mathrm{TE}} = (n_3^2/n_x^2 - u^2)^{1/2}$. The far-field radiation power spectrum per unit area from the dipole with respect to viewing angle θ_3 is described according to the photon spectrum $s(\lambda)$ as

$$I(\theta_3,\lambda) = \left[\begin{array}{l} (1-\alpha)\cdot P_0^{\perp}\cdot p_{\mathrm{out}}^{\prime\perp,\mathrm{TM}}(\theta_3,\lambda) \\ +\alpha\cdot P_0^{\parallel}\cdot\left\{ p_{\mathrm{out}}^{\prime\parallel,\mathrm{TM}}(\theta_3,\lambda) + p_{\mathrm{out}}^{\prime\parallel,\mathrm{TE}}(\theta_3,\lambda) \right\} \end{array} \right]\cdot s(\lambda). \quad (2.27)$$

2.2.4 Experimental

A mixed layer of 4,4′,4″-tris(carbazol-9-yl)-triphenylamine [TCTA] and bis-4,6-(3,5-di-3-pyridylphenyl)-2-methylpyrimidine [B3PyMPM] was used as a dielectric birefringent medium and a phosphorescent dye of bis(2-phenylpyri dine)iridium(III)(2,2,6,6-tetramethylheptane-3,5-diketonate) [Ir(ppy)$_2$tmd] [35, 36] doped in the mixed layer was used as an emitter. The mixed layer of TCTA and B3PyMPM was reported as an exciplex-forming host in OLEDs [6, 36–40] that enables effective energy transfer to the Ir(ppy)$_2$tmd. Films and OLEDs were fabricated using thermal evaporation in vacuum. Figure 2.2 shows the structure of the OLEDs with thick TCTA:B3PyMPM layers. The structure of the OLEDs consists of glass substrate/ITO (70 nm)/MoO$_3$ (1 nm)/TCTA (10 nm)/TCTA:B3PyMPM:Ir(ppy)$_2$tmd (45.8:45.8:8.4 mol%, 125 nm)/B3PMYPM (10 nm)/LiF (1 nm)/Al (100 nm). The doping region of Ir(ppy)$_2$tmd was varied for different emission zones in the same device structure. Four kinds of OLEDs were fabricated, having 20 nm-thick EMLs located 40, 55, 70, and 110 nm from the Al

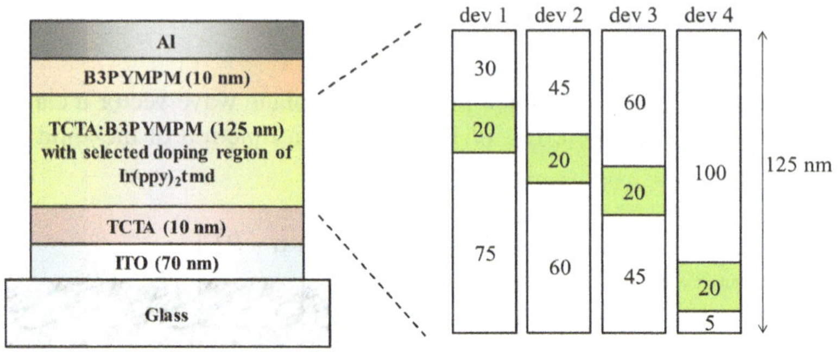

Fig. 2.2 Structures of OLEDs with a mixed host of TCTA: B3PyMPM with different doping regions

cathode and are referred to device 1, 2, 3, and 4, respectively. Thin MoO_3 and LiF layers were used for efficient hole and electron injection, respectively.

Variable angle spectroscopic ellipsometry (J. A. Woolam M-2000 spectroscopic ellipsometer) was used to analyze the molecular orientation and optical constants (refractive index n and extinction coefficient k) of the TCTA: B3PyMPM film deposited on the pre-cleaned silicon substrate. Optical constants were analyzed using J. A. Woolam Complete EASE software. Analysis of the optical constants was initiated using the Cauchy model at the transparent region, expanded to the whole region by the B-spline model satisfying the Kramers-Krönig consistency, and completed by inserting Gaussian oscillators into the result of the B-spline model. The uniaxial model was applied using separate analyses for the ordinary and the extraordinary axes. An angle-dependent PL analysis [23] was applied to determine the orientation of the transition dipole moment of $Ir(ppy)_2tmd$ in the TCTA: B3PyMPM host. We analyzed the dipole orientation with a 30 nm thick film of the $Ir(ppy)_2tmd$-doped TCTA: B3PyMPM layer deposited onto a fused silica substrate. The substrate was attached to a half-cylinder lens made of fused silica and fixed on a programmed rotation stage. The molecules in the film were excited by a He-Cd laser (325 nm, CW) and angle-dependent intensity profiles of the PL escaping through the lens were measured using the Maya2000 fiber spectrometer (Ocean Optics Inc.). For the analysis, TM-polarized light was selected using a linear polarizer.

The current-voltage-luminescence characteristics of the OLEDs were analyzed by the Keithley 2400 and the SpectraScan PR 650 (Photo Research). The angle-dependent emission spectra of the devices were measured using an Ocean Optics S2000 fiber optic spectrometer with constant current for angles ranging from 0° to 85°. Measurements were performed automatically using the programmed rotation stage. The EQEs of the OLEDs were obtained by calculating the ratio between the numbers of emitted photons and injected electrons. The calculation is performed by first obtaining the emission spectrum at a normal direction, assuming Lambertian distribution, and calibrated by considering the angle-dependent emission distribution of the OLEDs.

2.3 Results and Discussion

2.3.1 Optical Birefringence and the Dipole Orientation

The optical constants of the TCTA:B3PyMPM layer displayed negative birefringence ($n_o > n_e$) as shown by Fig. 2.3a, indicating the horizontal orientation of the molecules in the mixed layer. The difference between the ordinary and extraordinary refractive index n is ~0.2 in the visible region. Planar-shaped B3PyMPM molecules with hydrogen bonds result in a horizontal molecular orientation even in the vacuum-deposited organic film [41, 42]. Although they were mixed with the TCTA molecules, the horizontally preferred molecular orientation was observed in the mixed film.

Figure 2.3b shows the angular emission intensity profile of the TM wave from the film at the wavelength $\lambda = 520$ nm corresponding to the PL maximum. Optical simulations for the far-field emission of the TM wave from the film to the semi-infinite fused silica substrate was performed using Eq. (4.27) to determine the dipole orientation of the emitter, with an assumption that the molecules in the thin layer are excited uniformly throughout the layer. The horizontal dipole ratio (α) was taken as a fitting parameter for the experimental data. The experimental data are in good agreement with the theoretical prediction using $\alpha = 0.74$ across the entire emission angle when the birefringence of the emitting layer was considered. In contrast, the theoretical fittings under the assumption of an isotropic medium with an ordinary refractive index fit only part of the experimental data, i.e., over 40°. In other words, the accuracy of the predicted emission dipole orientation is significantly improved

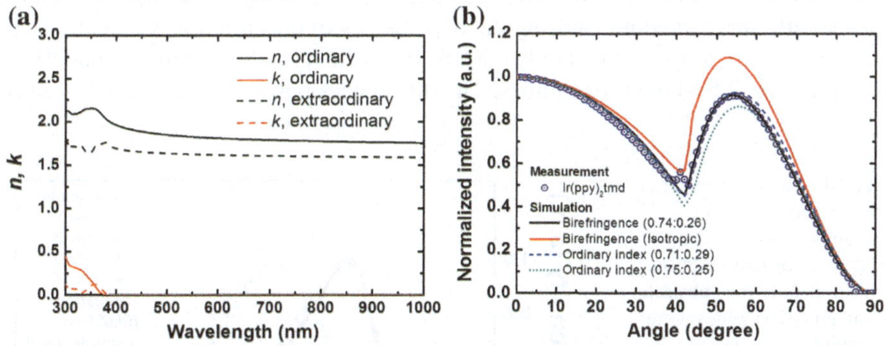

Fig. 2.3 **a** Optical constants of the TCTA: B3PyMPM layer. **b** Angle-dependent PL intensity profile of the Ir(ppy)$_2$tmd in the TCTA: B3PyMPM film at $\lambda = 520$ nm (circle) is compared with the theoretical values for an isotropic dipole orientation (red solid line) and for the emission dipoles with $\alpha = 0.74$ (black solid line) when considering a birefringent medium, and for $\alpha = 0.71$ (blue dashed line) and $\alpha = 0.75$ (green dotted line) without accounting for the birefringence of the emitting layer. Theoretical profiles calculated under the assumption of an isotropic medium fit only part of the experimentally obtained intensity profile

by accounting for the birefringence of the emission layer. The unexpected peak at around 40° in the measurement comes from reflection of the encapsulation glass at the opposite side of the substrate.

2.3.2 *Emission Spectra of OLEDs*

The emission spectra of the four different OLEDs along the normal direction with respect to the substrate are shown in Fig. 2.4. The emission spectra exhibit more pronounced longer-wavelength vibronic peaks as the distance between the doping region and the cathode increases. Emission spectra of the OLEDs were used to predict the location of the emission zone in the devices by fitting the far field radiation with the location of the emission zone as a parameter, where the emission zone geometry is assumed to be that of a sheet. The mean emission zones of the four devices were determined to be located at 50 nm (device 1), 60 nm (device 2), 75 nm (device 3), and 120 nm (device 4) from the cathode, all of which are located in the doped regions of the devices. Note that the optical birefringence is not effective in this calculation because only the ordinary refractive index of the medium affects light propagating in the direction normal to the substrate. However, consideration of the birefringence is important in determining the far-field radiation at other angles.

The measured angle-dependent EL spectra of devices 1, 2, 3, and 4 between 0° and 80° are depicted in Fig. 2.5. As the emission zone is located far from the cathode, the angular emission spectra were broadened and the intensity at high angles increased. The calculated far-field radiant spectra that include the effects of birefringence (solid lines) agreed with the experimental results in intensity, resonance wavelength, spectral width, and angular dependency of the radiation. Calculations of the angular radiant spectra that only consider an ordinary index (isotropic model) are shown in Fig. 2.5 as the broken lines. The differences between them are not signifi-

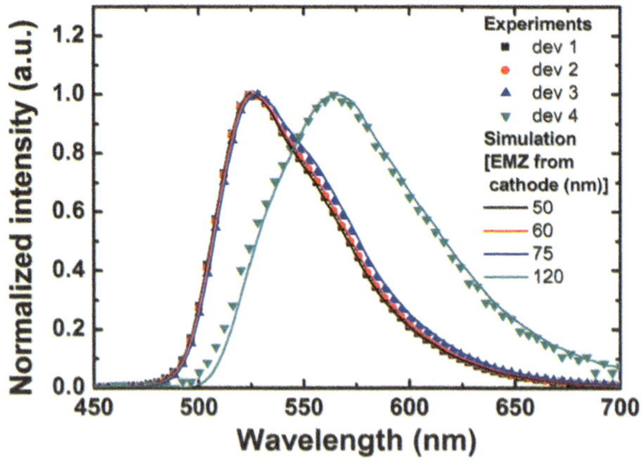

Fig. 2.4 Emission spectra of OLEDs with different locations in the emission layer at a normal direction with respect to the substrate. Mean emission zones were determined via optical simulation based on the emission spectra

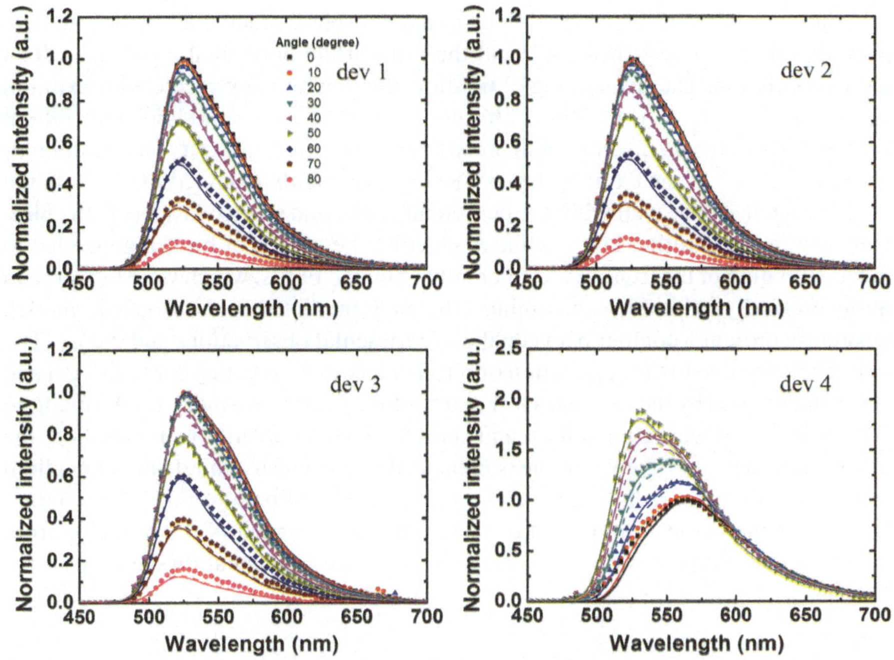

Fig. 2.5 Angle-dependent EL spectra of devices 1, 2, 3, and 4 (points) compared to the calculated far-field radiation spectra when considering the birefringence (solid lines) and for an ordinary index of refraction (broken lines)

cant in the case of device 1, 2, and 3, but are significant in the device 4, which has a gap in the resonance wavelength from the PL maximum of the $Ir(ppy)_2tmd$. The calculation including birefringence yielded a larger radiant intensity at high angles than that for an ordinary index, and better explains the far-field emission characteristics of the device.

2.3.3 *Efficiency of OLEDs*

Current density-voltage-luminance (*J-V-L*) characteristics of the four kinds of OLEDs are depicted in Fig. 2.6a. They show general diode characteristics with consistent turn-on voltages of 2.7 V and low leakage current. EQEs of the four kinds of OLEDs versus current density are shown in Fig. 2.6b. The maximum EQEs of devices were 28.0, 28.8, 27.9, and 11.6% for the device 1, 2, 3, and 4, respectively. Their maximum EQEs were obtained at the low current density, indicating electrical losses at high current densities. Calculation of the maximum EQEs of OLEDs as a function of the distance to the emission zone from the cathode as estimated using Eq. (2.23) is displayed in Fig. 2.6c. We chose the electrical balance factor $\gamma = 1$ (no electrical loss), the spin statistics factor $\chi = 1$ due to the phosphorescent emitter, the radiative quantum efficiency $q = 0.96$ [36], and the ratio of the horizontal dipole $\alpha = 0.74$ in the calculation. The lines in Fig. 2.6c show the theoretically predicted maximum EQEs as a function of the location of the emission zone with (solid line) and without (dashed line, isotropic medium with an ordinary refractive index) a consideration of the birefringence of the emitting layer. The experimentally obtained EQEs from the OLEDs matched very well with the theoretical prediction when considering the birefringence in the emitting layer, while neglecting the effects of birefringence led to underestimation of the EQE. Experimentally obtained EQEs, which were higher than the theoretically achievable maximum efficiency predicted in the isotropic model, apparently present a conflict between the experimental observations and theory. The high efficiency due to the application of birefringence resolves this contradiction and can be understood by the fact that a low extraordinary refractive index for the medium reduces losses of waveguided light and surface plasmon polaritons. In addition, the radiation from the horizontal dipole is enhanced in the negative birefringent medium because the intrinsic radiation power of the horizontal dipole in Eq. (2.2) is larger than that of the vertical dipole in Eq. (2.1). The enhancement of the radiation from the horizontal dipole is similar to the increase of the horizontal dipole ratio in the OLEDs.

2.3.4 Conclusion

We have presented an optical model for emission in a birefringent medium and have validated the theoretical model by applying it to a thin film and OLEDs. We have demonstrated that the optical birefringence affects not only far-field radiation properties such as emission spectra from the film and OLEDs, but also the efficiency of the OLEDs. The emitting dipole orientation in the birefringent medium has been successfully analyzed from the far-field radiation pattern of a thin film. In addition, the birefringent model has provided a precise analysis of angle-dependent EL spectra and EQEs of OLEDs with the determined emitting dipole orientation.

2.3.5 Appendix

The Fresnel's reflection and transmission coefficients for TM and TE polarized light traveling from layer a to b are expressed in terms of the wave vector components and the refractive index of layers a and b. If the layer a has an anisotropic refractive index tensor $\tilde{\tilde{n}}$

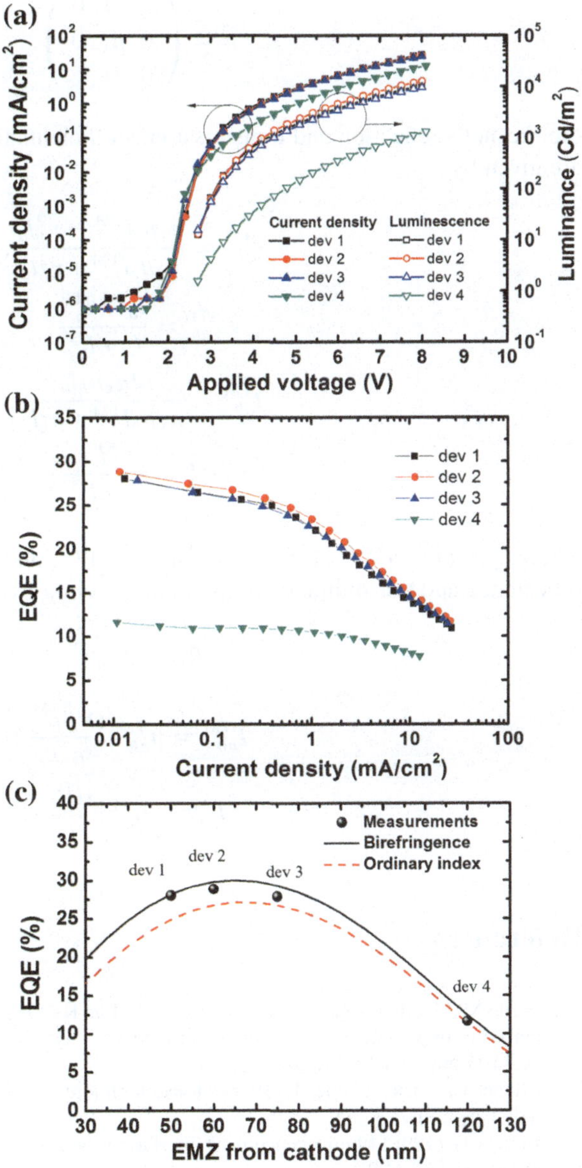

Fig. 2.6 **a** Current density-voltage-luminance curves of devices 1, 2, 3, and 4. **b** EQE-current density characteristics of devices 1, 2, 3, and 4. **c** The experimentally obtained maximum EQEs (filled circles) of the OLEDs are compared with theoretical calculations performed with (solid lines) and without (dashed lines) considering the birefringence of the emitting layer, under the assumption of no electrical loss in the device. The experimental results are in excellent agreement with the theoretical predictions for emission dipoles in the birefringent medium, and are higher than the values predicted for an isotropic medium

$$\tilde{\tilde{n}} = \begin{pmatrix} n_x & 0 & 0 \\ 0 & n_x & 0 \\ 0 & 0 & n_z \end{pmatrix},$$

the Fresnel's reflection and transmission coefficients for TM and TE polarized light are given by

$$r_{ab}^{\mathrm{TM}} = \frac{n_x n_z l_b^{\mathrm{TM}} - n_b^2 l_a}{n_x n_z l_b^{\mathrm{TM}} + n_b^2 l_a}, \tag{2.28}$$

$$r_{ab}^{\mathrm{TE}} = \frac{l_a - l_b^{\mathrm{TE}}}{l_a + l_b^{\mathrm{TE}}}, \tag{2.29}$$

$$t_{ab}^{\mathrm{TM}} = \frac{2 n_x n_b l_a}{n_x n_b l_b^{\mathrm{TM}} + n_b^2 l_a}, \tag{2.30}$$

$$t_{ab}^{\mathrm{TE}} = \frac{2 l_a}{l_a + l_b^{\mathrm{TE}}}, \tag{2.31}$$

where $l_a = (1 - u^2)^{1/2}$, $l_b^{TM} = (n_b^2/n_z^2 - u^2)^{1/2}$, and $l_b^{TE} = (n_b^2/n_x^2 - u^2)^{1/2}$. The reflectance and transmittance of the light traveling from layer a to b are given by

$$R_{ab} = |r_{ab}|^2, \tag{2.32}$$

$$T_{ab}^{\mathrm{TM}} = \left|t_{ab}^{\mathrm{TM}}\right|^2 \frac{n_z l_b^{\mathrm{TM}}}{n_x l_a}, \tag{2.33}$$

$$T_{ab}^{\mathrm{TE}} = \left|t_{ab}^{\mathrm{TE}}\right|^2 \frac{l_b^{\mathrm{TE}}}{l_a}. \tag{2.34}$$

References

1. Furno M, Meerheim R, Hofmann S, Lüssem B, Leo K (2012) Efficiency and rate of spontaneous emission in organic electroluminescent devices. Phys Rev B 85(11):115205. https://doi.org/10.1103/physrevb.85.115205
2. Chance R, Prock A, Silbey R (1978) Molecular fluorescence and energy transfer near interfaces. Adv Chem Phys 37:65
3. Barnes W (1998) Fluorescence near interfaces: the role of photonic mode density. J Modern Optics 45:661–699
4. Neyts KA (1998) Simulation of light emission from thin-film microcavities. JOSA A 15:962–971
5. Wasey J, Safonov A, Samuel I, Barnes W (2000) Effects of dipole orientation and birefringence on the optical emission from thin films. Optics Commun 183:109–121
6. Wasey JAE, Safonov A, Samuel IDW, Barnes WL (2001) Efficiency of radiative emission from thin films of a light-emitting conjugated polymer. Phys Re B 64(20):205201. https://doi.org/10.1103/physrevb.64.205201

7. Lin C-L, Cho T-Y, Chang C-H, Wu C-C (2006) Enhancing light outcoupling of organic light-emitting devices by locating emitters around the second antinode of the reflective metal electrode. Appl Phys Lett 88:081114
8. Lin C-L, Chang H-C, Tien K-C, Wu C-C (2007) Influences of resonant wavelengths on performances of microcavity organic light-emitting devices. Appl Phys Lett 90:071111
9. Kim, J-B, Lee J-H, Moon C-K Kim J-J (2014) Highly efficient inverted top emitting organic light emitting diodes using a transparent top electrode with color stability on viewing angle. Appl Phys Lett 104: 31–33
10. Bulović V et al (1998) Weak microcavity effects in organic light-emitting devices. Phys Rev B 58:3730
11. Wasey J, Safonov A, Samuel I, Barnes W (2000) Effects of dipole orientation and birefringence on the optical emission from thin films. Optics Commun 183:109–121
12. Kim SY et al (2013) Organic Light-Emitting diodes with 30% external quantum efficiency based on a horizontally oriented emitter. Adv Func Mater 23:3896–3900
13. Shin H et al (2014) Blue phosphorescent organic light-emitting diodes using an exciplex forming co-host with the external quantum efficiency of theoretical limit. Adv Mater 26:4730–4734
14. Lu M-H, Sturm J (2002) Optimization of external coupling and light emission in organic light-emitting devices: modeling and experiment. J Appl Phys 91:595–604
15. Smith LH, Wasey JA, Samuel ID, Barnes WL (2005) Light out-coupling efficiencies of organic light-emitting diode structures and the effect of photoluminescence quantum yield. Adv Func Mater 15:1839–1844
16. Nowy S, Krummacher BC, Frischeisen J, Reinke NA, Brütting W (2008) Light extraction and optical loss mechanisms in organic light-emitting diodes: influence of the emitter quantum efficiency. J Appl Phys 104:123109
17. Meerheim R, Furno M, Hofmann S, Lüssem B, Leo K (2010) Quantification of energy loss mechanisms in organic light-emitting diodes. Appl Phys Lett 97:275
18. Kim S-Y, Kim J-J (2010) Outcoupling efficiency of organic light emitting diodes and the effect of ITO thickness. Org Electron 11:1010–1015
19. Lee J-H, Lee S, Kim J-B, Jang J, Kim J-J (2012) A high performance transparent inverted organic light emitting diode with 1, 4, 5, 8, 9, 11-hexaazatriphenylenehexacarbonitrile as an organic buffer layer. J Mater Chem 22:15262–15266
20. Kim JB, Lee JH, Moon CK, Kim SY, Kim JJ (2013) Highly enhanced light extraction from surface plasmonic loss minimized organic light-emitting diodes. Adv Mater 25:3571–3577
21. Kim KH, Moon CK, Lee JH, Kim SY, Kim JJ (2014) Highly efficient organic light-emitting diodes with phosphorescent emitters having high quantum yield and horizontal orientation of transition dipole moments. Adv Mater 26:3844–3847
22. Sun JW et al (2014) A fluorescent organic light-emitting diode with 30% external quantum efficiency. Adv Mater 26:5684–5688
23. Frischeisen J, Yokoyama D, Adachi C, Brütting W (2010) Determination of molecular dipole orientation in doped fluorescent organic thin films by photoluminescence measurements. Appl Phys Lett 96:073302. https://doi.org/10.1063/1.3309705
24. Flämmich M et al (2011) Oriented phosphorescent emitters boost OLED efficiency. Org Electron 12:1663–1668
25. Liehm P et al (2012) Comparing the emissive dipole orientation of two similar phosphorescent green emitter molecules in highly efficient organic light-emitting diodes. Appl Phys Lett 101:253304
26. Flämmich M et al (2010) Orientation of emissive dipoles in OLEDs: quantitative in situ analysis. Org Electron 11:1039–1046
27. Schmidt TD et al (2011) Evidence for non-isotropic emitter orientation in a red phosphorescent organic light-emitting diode and its implications for determining the emitter's radiative quantum efficiency. Appl Phys Lett 99:225
28. Penninck L, Steinbacher F, Krause R, Neyts K (2012) Determining emissive dipole orientation in organic light emitting devices by decay time measurement. Org Electron 13:3079–3084

29. Lin H-W et al (2004) Anisotropic optical properties and molecular orientation in vacuum-deposited ter (9, 9-diarylfluorene) s thin films using spectroscopic ellipsometry. J Appl Phys 95:881–886
30. Yokoyama D, Sakaguchi A, Suzuki M, Adachi C (2008) Horizontal molecular orientation in vacuum-deposited organic amorphous films of hole and electron transport materials. Appl Phys Lett 93:394
31. Yokoyama D (2011) Molecular orientation in small-molecule organic light-emitting diodes. J Mater Chem 21:19187–19202
32. Yokoyama D, Nakayama KI, Otani T, Kido J (2012) Wide-range refractive index control of organic semiconductor films toward advanced optical design of organic optoelectronic devices. Adv Mater 24(47):6368–6373
33. Penninck L, De Visschere P, Beeckman J, Neyts K (2011) Dipole radiation within one-dimensional anisotropic microcavities: a simulation method. Opt Express 19:18558–18576
34. Purcell EM (1995) Spontaneous emission probabilities at radio frequencies. In: Confined Electrons and Photons. Springer, Boston, MA, pp 839–839
35. Lamansky S et al (2001) Highly phosphorescent bis-cyclometalated iridium complexes: synthesis, photophysical characterization, and use in organic light emitting diodes. J Am Chem Soc 123:4304–4312
36. Kim KH, Moon CK, Lee JH, Kim SY, Kim JJ (2014) Highly efficient organic light-emitting diodes with phosphorescent emitters having high quantum yield and horizontal orientation of transition dipole moments. Adv Mater 26:3844–3847. https://doi.org/10.1002/adma.201305733
37. Kim S-Y et al (2013) Organic light-emitting diodes with 30% external quantum efficiency based on a horizontally oriented emitter. Adv Func Mater 23:3896–3900. https://doi.org/10.1002/adfm.201300104
38. Park YS et al (2013) Exciplex-forming co-host for organic light-emitting diodes with ultimate efficiency. Adv Func Mater 23:4914–4920
39. Lee JH, Lee S, Yoo SJ, Kim KH, Kim JJ (2014) Langevin and trap-assisted recombination in phosphorescent organic light emitting diodes. Adv Func Mater 24:4681–4688
40. Lee S, Kim KH, Limbach D, Park YS, Kim JJ (2013) Low roll-off and high efficiency orange organic light emitting diodes with controlled co-doping of green and red phosphorescent dopants in an exciplex forming co-host. Adv Func Mater 23:4105–4110
41. Yokoyama D (2011) Molecular orientation in small-molecule organic light-emitting diodes. J Mater Chem 21:19187. https://doi.org/10.1039/c1jm13417e
42. Sasabe H et al (2011) Influence of substituted pyridine rings on physical properties and electron mobilities of 2-methylpyrimidine skeleton-based electron transporters. Adv Func Mater 21:336–342

Chapter 3
The Orientation of Ir Complexes Doped in Organic Amorphous Layers

3.1 Influence of Host Materials on the Emitting Dipole Orientation of Ir Complexes

3.1.1 Introduction

Molecular orientation in organic semiconductor is an important factor influencing electrical and optical properties [1]. In OLEDs, emitting dipole orientation (EDO) is directly related to the outcoupling efficiency of the light. Therefore, employing horizontally oriented emitter is one of the effective methods to enhancing the outcoupling efficiency of OLEDs [2–13]. Recent researches have focused on molecular structure of emitters [5, 8, 10, 12, 13] or changing deposition temperature to understand the preferred EDO of emitters doped in host layers [14, 15].

Recently, Kim et al. [5] reported that two conditions must be satisfied to have a preferred orientation of the transition dipole moments in the dye-doped co-host films: (1) the molecule should have triplet transition dipole moments preferentially oriented along a specific direction and not have a combination of transition dipole moments with various orientations, and (2) the molecule itself should have a preferred orientation with respect to substrate. They proposed based on quantum calculation that the heteroleptic molecular structure of the Ir (III) complexes and strong Coulombic interactions between the dopant and host molecules are the driving forces of the preferred EDO. In this regard, not only the structure of the emitting molecules but also the organic host molecules must influence the EDO. The different degrees of EDO for certain phosphorescent dyes have been reported in different hosts in literature. Unfortunately, however, there are few reports on the systematic investigation of the host effect on the EDO.

In this chapter, we report that the EDO of phosphorescent dyes is strongly related to the intermolecular interaction between the host and dye molecules influenced by the molecular structures of the hosts and processing conditions. The EDO of heteroleptic Ir complexes changes preferentially from horizontal to isotropic and

C.-K. Moon, *Molecular Orientation and Emission Characteristics of Ir Complexes and Exciplex in Organic Thin Films*, Springer Theses,
https://doi.org/10.1007/978-981-13-6055-8_3

vertical direction with respect to the substrate depending on the host molecules deposited at room temperature. Solution processed films in a polymer resulted in random orientation of the dyes. It could be understood based on the intermolecular binding forces calculated by quantum chemical calculation.

3.1.2 Results and Discussion

Four different host molecules and two kinds of phosphorescent Ir-complexes were prepared to investigate the influence of host molecules and thereby to explore the effects of the intermolecular interaction between host and dopant molecules on the EDO of dopants. The chemical structures of host molecules and Ir-complexes are shown in Fig. 3.1. We used *N*,*N*′-dicarbazolyl-4-4′-biphenyl [CBP], bis-4,6-(3,5-di-3-pyridylphenyl)-2-methylpyrimidine [B3PyMPM], and 1,7-bis(triphenylsilyl)benzene [UGH-2] for hosts containing aromatic rings, and poly(methyl methacrylate) [PMMA] as an aliphatic host. In terms of the molecular configuration, CBP and B3PyMPM have linear or planar molecular structures composed of sp^2 hybridizations but UGH-2 has a bulky symmetric molecular structure with sp^3 hybridizations of silicon atoms. In terms of the molecular orientation in film state, CBP and UGH-2 layers are isotropic with the random orientation of the molecules at room temperature [13], but the B3PyMPM layer has a horizontally preferred molecular orientation parallel to substrate [15]. A green emitting homoleptic complex of tris(2-phenylpyridine)iridium(III) [$Ir(ppy)_3$] and a heteroleptic complex of bis(2-phenylpyridine)iridium(III)(2,2,6,6-tetramethylheptane-3,5-diketonate) [$Ir(ppy)_2tmd$] [8, 16] were selected as the phosphorescent dyes.

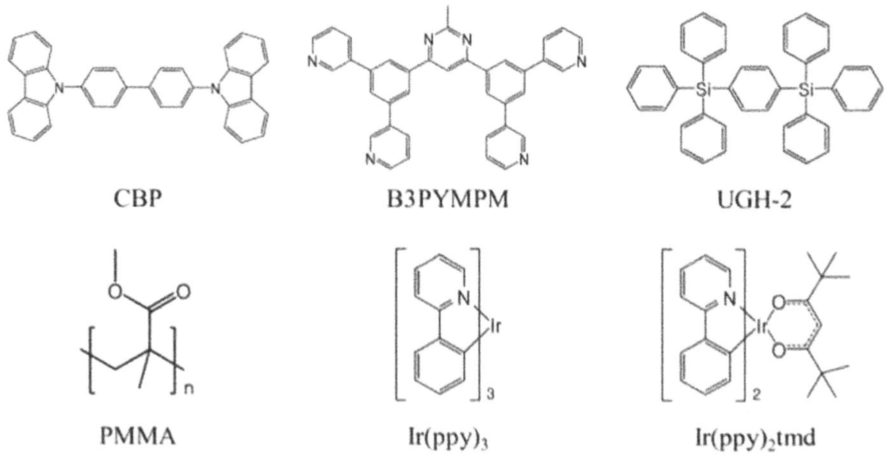

Fig. 3.1 Molecular structure of host molecules (CBP, B3PyMPM, UGH-2), and green phosphorescent iridium complexes ($Ir(ppy)_3$, $Ir(ppy)_2tmd$) used in this study

The 30 nm thick films of CBP, B3PyMPM, and UGH-2 with 10 mol% $Ir(ppy)_3$ and $Ir(ppy)_2tmd$, respectively, were fabricated onto the fused silica substrates by vacuum deposition. In addition, 22 nm thick films of PMMA doped with 10 wt% $Ir(ppy)_3$ and $Ir(ppy)_2tmd$, respectively, were fabricated onto the fused silica substrate by spin coating using chloroform as solvent. The distribution of EDO in the films were determined by fitting the measured angle-dependent PL emission spectra of the p-polarized light from the films using the optical simulation for far-field emission intensity from the films [17]. Optical birefringence of the B3PyMPM layer resulting from the preferred horizontal orientation of the molecules in the film state was considered in the analysis [18].

The EDOs of $Ir(ppy)_3$ and $Ir(ppy)_2tmd$ doped in the four different hosts were extracted from the angle-dependent PL intensities at the wavelength of 520 nm shown in Fig. 3.2 as open squares and circles, respectively. Black and red lines in Fig. 3.2 are the theoretical fittings using the classical dipole model for the far-field emission of the films with specific dipole distributions. All emission patterns of the Ir complexes are in good agreements with the simulation. Fittings of the angle-dependent PL intensity profiles for the entire emission spectral range are given in Fig. 3.3. The extracted EDOs are summarized in Table 3.1. These results clearly show that different hosts result in significantly different EDOs for the dopants. For instance, $Ir(ppy)_3$ has almost isotropic EDOs in CBP and B3PyMPM, whereas $Ir(ppy)_2tmd$ has preferred horizontal EDOs in the hosts. One can note that the horizontal to vertical dipole ratio is 0.67:0.33 for the isotropic (random) dipole orientation. Both hosts result in similar EDOs for each dopant. In contrast, UGH-2 orients the dopants to have rather vertical EDOs, and the solution processed PMMA induces random orientation for the both dopants.

The isotropic EDOs of $Ir(ppy)_3$ in CBP, B3PyMPM, and PMMA can be easily understood from the three mutually orthogonal triplet transition dipole moments along the directions from the iridium core to the three C^N ligands with 3-fold rotational symmetry (Fig. 3.4). Combination of the three equivalent dipole moments reduces the preferred EDO to a specific direction, resulting in almost isotropic EDO in the hosts. In contrast, $Ir(ppy)_2tmd$ has two nearly parallel triplet transition dipole moments along the directions from the iridium core to the two C^N ligands with 2-fold symmetry 10. Therefore, $Ir(ppy)_2tmd$ can have larger anisotropy of EDO than $Ir(ppy)_3$ if the molecules align themselves with respect to the substrate. Much less preferred EDO of $Ir(ppy)_3$ than $Ir(ppy)_2tmd$ in UGH-2 can be understood by the same token. Strong dependence of the EDO of $Ir(ppy)_2tmd$ on host molecules clearly indicates that the iridium complex–host interaction plays an important role in orienting the emission dipoles.

We have performed the DFT calculation to investigate the binding energy and geometries between the host molecules and the dyes using the method as described before [19]. The calculated host–host, host–dopant, and dopant–dopant binding energies are summarized in Table 3.2. Both phosphorescent dyes have large binding energies of about 30 kcal/mol (125 kJ/mol) with the CBP and B3PyMPM which are more than 10 times larger than the host–host and dopant–dopant binding energies.

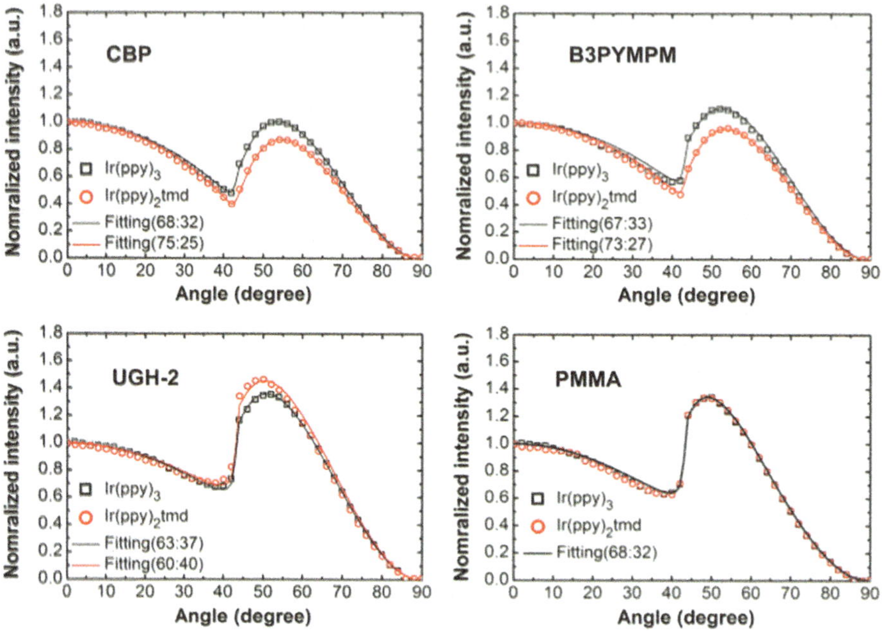

Fig. 3.2 Angle-dependent PL profiles of $Ir(ppy)_3$ (black open squares) and $Ir(ppy)_2tmd$ (red open circles) in CBP, B3PyMPM, UGH-2, and PMMA, respectively, at a wavelength $\lambda = 520$ nm with fittings by the optical simulation (black and red lines). Horizontal dipole ratio is 0.67 and vertical dipole ratio is 0.33 for an isotropic dipole orientation (described as 67:33 in the figure)

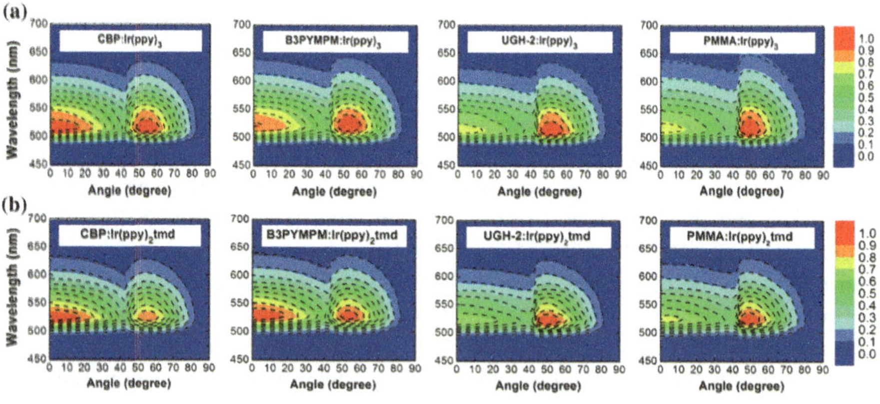

Fig. 3.3 Full emission spectra of the films containing **a** $Ir(ppy)_3$ and **b** $Ir(ppy)_2tmd$ (colored surfaces) with their fittings by the optical simulation (dashed lines) employing the determined dipole orientation

Table 3.1 The horizontal to vertical emitting dipole orientations of $Ir(ppy)_3$ and $Ir(ppy)_2tmd$ in various host molecules

Ir complex	CBP	B3PYMPM	UGH-2	PMMA
$Ir(ppy)_3$	0.68:0.32	0.67:0.33	0.63:0.37	0.68:0.32
$Ir(ppy)_2tmd$ (preferred orientation)	0.75:0.25 (horizontal)	0.73:0.27 (horizontal)	0.60:0.40 (horizontal)	0.68:0.32 (isotropic)

Figure 3.5a and b shows the optimized binding geometries of $Ir(ppy)_2tmd$–CBP and $Ir(ppy)_2tmd$–B3PyMPM, respectively. The binding geometries are linear, and their long axes are perpendicular to the C_2 axis of $Ir(ppy)_2tmd$. Many papers reported

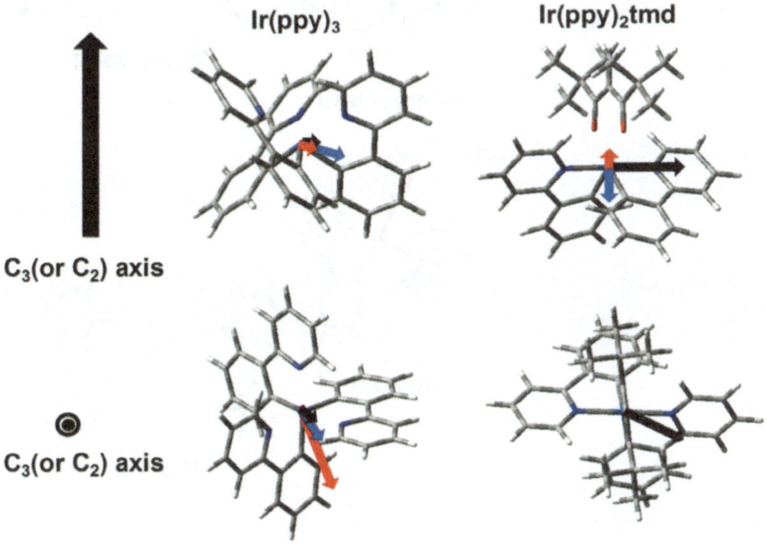

Fig. 3.4 Triplet transition dipole moments in $Ir(ppy)_3$ and $Ir(ppy)_2tmd$ molecules. Black, blue, red arrows indicate the triplet transition dipole moment of Tx, Ty, Tz sublevels, respectively

Table 3.2 Calculated intermolecular binding energies between two molecules

	$Ir(ppy)_3$	$Ir(ppy)_2tmd$	CBP	B3PYMPM	UGH-2
$Ir(ppy)_3$	−2.70		−30.64	−28.13	6.26 (unstable)
$Ir(ppy)_2tmd$		−2.26	−31.26	−30.95	5.95 (unstable)
CBP			−3.31		
B3PYMPM				−2.02	
UGH-2					−0.44

The negative sign indicates attractive energy and the positive sign indicates repulsive energy in kcal/mol

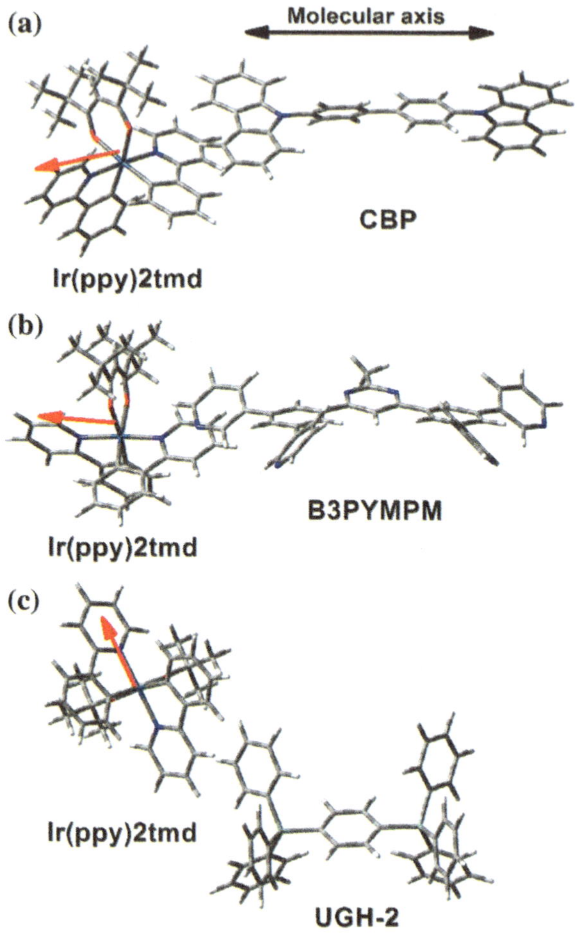

Fig. 3.5 Optimized binding geometries of $Ir(ppy)_2tmd$ with **a** CBP, **b** B3PyMPM, and **c** UGH-2 molecules. The black arrow shows long axis of the molecules, and red arrows indicate the orientation of the transition dipole moment

that organic molecules possessing linear or planar molecular structures have preferred horizontal orientation even in the vacuum deposited films [1, 13, 14]. In the same manner, the linear binding geometries of the dopant–host molecules with the large binding energy are expected to induce the bound molecules to align horizontally along the substrate during vacuum deposition, and thereby the transition dipole moment of $Ir(ppy)_2tmd$ is oriented preferentially toward the horizontal direction. It is interesting to note that the randomly oriented CBP host results in a little higher horizontal dipole ratio of $Ir(ppy)_2tmd$ than B3PyMPM, implying that EDO is not controlled by the orientation of host molecules but by the binding geometry of host–dopant molecules.

In contrast to CBP and B3PyMPM, the calculated binding energies between UGH-2 and $Ir(ppy)_2tmd$ or $Ir(ppy)_3$ are positive, implying that the bindings are thermodynamically unstable (see Table 3.2). However, the values are small enough not to prevent the hosts and dopant molecules from the formation of the condensed films.

Figure 3.5c shows the optimum binding geometry of $Ir(ppy)_2tmd$ and UGH-2 with the locally minimized energy state. The C^N ligand of the $Ir(ppy)_2tmd$ is apart from the UGH-2 molecule or the UGH-2 film surface due to the electro-repulsion forming nonlinear optimum geometry to induce the vertical EDO of the dyes.

There is no clear explanation at this moment on the reason why both dyes have isotropic orientation in PMMA. One plausible reason is the weak intermolecular interaction expected between the dopant and the aliphatic chain to randomize the orientation of the dyes by entropic energy. Another possible reason is the thermal history of the films which had experienced the annealing process at the temperature over the glass transition temperature. Further study is required to fully understand the reason. We could not perform the quantum chemical calculation for PMMA because of the limitation of the computational power.

3.1.3 Conclusion

In summary, the EDO of phosphorescent dyes doped in organic thin films is significantly influenced not only by the molecular structure of the dyes but also by the host molecules. *Fac*-$Ir(ppy)_3$, a homoleptic iridium complex, has almost isotropic EDO in any kinds of host molecules because of the combination of its three mutually orthogonal triplet transition dipole moments. However, $Ir(ppy)_2tmd$, a heteroleptic iridium complex, can have preferred EDO in host layers due to the two parallel triplet transition dipole moments. The EDO of the heteroleptic Ir complex varies from horizontal to isotropic or even to vertical direction depending on host molecules. Interestingly enough, the orientation of host molecules does not significantly influence the EDO of the emitter. The DFT calculation indicates that the horizontal EDO is induced when the linear binding geometry of the host–dopant molecules is parallel to the transition dipole moment of $Ir(ppy)_2tmd$ with the large binding energy. On the other hand, the vertical EDO is induced when they interact repulsively during the film deposition to form nonlinear binding geometry.

3.2 Unraveling the Orientation of Ir Complexes via Vacuum Deposition Simulation

3.2.1 Introduction

It is only in recent years has attention turned to the orientation of emitting dipoles of iridium-based phosphors, the most verified light emitting dyes with high PL quantum yield and variety of chromatic spectrum as doped in the emissive layers; probably because their iridium-centered spherical shape and the amorphous surrounding nature in the emissive layers are far from having strong molecular alignments. Recently, some heteroleptic Ir complexes exhibiting efficient electroluminescence in OLEDs are reported to possess preferred horizontal emitting dipole orientations (EDOs) [2–5, 20–22]. However, it was difficult to assert the reason why the spherical-shaped phosphors have a propensity toward preferred molecular alignment in the emissive layers. A few mechanisms have been proposed to explain the preferred molecular orientation of the Ir complexes doped in vacuum deposited organic semiconducting layers: (1) molecular aggregation of the dopants leading to randomizing their orientation by suppressing the intermolecular interaction between the dopant and host molecules [23], (2) strong intermolecular interactions between electro-positive sides of the dopant and the electro-negative host molecules promoting parallel alignment of the *N*-heterocycles of Ir-complexes by forming host-dopant-host pseudo-complex mainly participating in ^{3}MLCT transition [5, 24], and (3) π–π interactions between the dopant and host molecules on the organic surface bringing alignment of aliphatic ligands to the vacuum side [21, 25]. Currently, it is not very clear which mechanism most comprehensively describes the origin of the preferred EDO of the heteroleptic iridium phosphors. Moreover, the models are too oversimplified to describe the EDO values quantitatively, which depend on structures of the phosphors and host molecules [24]. Therefore, the molecular configurations and the interactions responsible for the EDOs of Ir-complexes should be established by atomic-scale simulation of the Ir complexes interacting with host molecules during film fabrication.

In this chapter, we carefully examine the vacuum deposition process of phosphors on organic layers using a combination of molecular dynamics (MD) simulations and quantum mechanical (QM) analyses. The triplet EDO of heteroleptic Ir complexes doped in organic layers are studied with systematic variations of the molecular structures of both host and dopant. Theoretical prediction of EDO from simulated deposition process reveals excellent quantitative agreement with experimental observations, reproducing the anisotropic molecular orientations of heteroleptic Ir complexes in the emissive layers. In-depth analysis indicates that the molecular orientation originates from the coupling of the cyclometalated main ligand participating in the optical transition with neighbor host molecules rather than from the alignment of aliphatic ancillary ligand toward vacuum. Close observation of the simulation results indicate that non-bonded energy has a critical influence on the molecular orientation during the deposition.

3.2.2 Results

Modeling of emitting dipole orientation

The simulation method for obtaining EDO of an emitter in the vacuum deposited layer is schematically illustrated in Fig. 3.6a. Firstly, the transition dipole moment (TDM) vector in the molecular frame ($\boldsymbol{m}_x$-, $\boldsymbol{m}_y$-, and $\boldsymbol{m}_z$-axes) was determined by QM calculations after optimization of molecular geometry. For iridium-based phosphors, spin-orbit coupled time-dependent density functional theory (SOC-TDDFT) was employed for the calculation of the triplet TDM vectors for phosphorescence. Secondly, vacuum deposition of the emitting molecules on organic surfaces was simulated using MD. Finally, the TDM vectors in the molecular axis in each frame of MD were transformed to the vectors in the laboratory axis ($\boldsymbol{n}_x$-, $\boldsymbol{n}_y$-, and $\boldsymbol{n}_z$-axes) by rotation matrix method (Fig. 3.6b). We determine φ_C and φ_L as the angle between $\boldsymbol{m}_z$ and $\boldsymbol{n}_z$ axes and the angle between the TDM vector of the emitter and $\boldsymbol{n}_z$ axis, representing the molecular orientation and the EDO against the vertical direction in the laboratory axis, respectively. The ratio of the horizontal (TDM_H) to the vertical transition dipole moment (TDM_V) follows the trigonometric relationship:

$$\mathrm{TDM_H} : \mathrm{TDM_V} = \mu_0^2 \sin^2 \varphi_\mathrm{L} : \mu_0^2 \cos^2 \varphi_\mathrm{L}, \tag{3.1}$$

where μ_0 is the magnitude of the dipole moment and squares of the components indicate the intensity of the transition (emission intensity). The EDO describes an average fraction of the horizontal and vertical dipole moment of whole emitters embedded in the emissive layer. An ensemble average of the horizontal dipole moment gives the fraction of horizontal emitting dipole moment in the emissive layer (Θ) as a parameter of the EDO by

$$\Theta = \langle \sin^2 \varphi_\mathrm{L} \rangle. \tag{3.2}$$

Details about the rotation matrix and the vector transformation are given in Methods section.

The deposition simulation was performed by dropping a target molecule onto organic substrates under vacuum followed by thermal equilibration at 300 K as shown in Fig. 3.6c. The simulations were performed using the Materials Science Suite (Version 2.2) released by Schrödinger Inc. [26]. Force field of OPLS_2005 [27] and periodic boundary conditions were used for the MD simulations. Preparation of the substrate had three steps of the MD simulation after locating 256 molecules in a grid: annealing at 500 K at first (NVT, 500 ps) followed by annealing at 300 K (NVT, 200 ps), and finally packing of the molecules at 300 K and 1 atm (NPT, 1000 ps for UGH-2 and CBP, and 5000 ps for TSPO1). Figure 3.7a shows the trajectories of the density and total potential energy of UGH-2, CBP, and TSPO1 substrates, respectively, at the step of packing. Density and potential energies were converged during the 1000 ps of simulation for the UGH-2 and CBP substrates and 5000 ps of simulation for the TSPO1 substrate. The NPT MD simulation produced amorphous solid densities of 1.04, 1.13, and 1.10 g/cm^3 for UGH-2, CBP, and TSPO1, respectively. Consider for example CBP whose simulated density of 1.13 g/cm^3 is in good agreement with the experimental value of 1.18 g/cm^3 [28]. Figure 3.7b exhibits the orientation of 256 molecular vectors of UGH-2, CBP, and TSPO1 indicated as red arrows, respectively, consisting the substrates. Their angular distributions were closed to the random distribution line so we concluded that the amorphous organic substrates were successfully prepared.

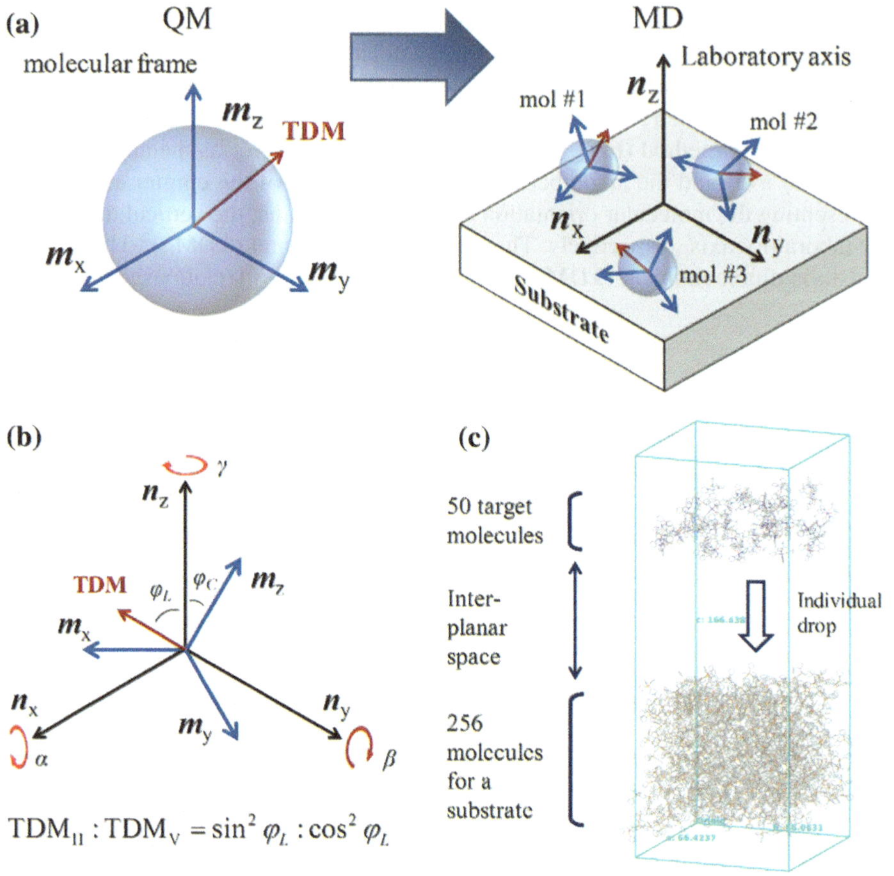

Fig. 3.6 Method for simulation of the EDO of emitters in vacuum-deposited layers. **a** Transfer of the TDM vectors (red arrow) in the molecular coordinates to the vectors of the molecules on the organic substrate during the vacuum deposition simulation. **b** Three rotation angles (α, β, and γ for the clockwise rotation to the n_x-, n_y-, and n_z-axes, respectively) were the orientation parameters of the molecules to correlate the molecular orientation to the laboratory axis. Angles between n_z axis and the TDM vector (φ_L) and the C_2 axis (φ_C) are obtained after the vector transformation. **c** A simulation box consisting of the substrate and target molecules. 50 target molecules were located above the substrate with 5.0 nm of inter-planar space dropped individually at 300 K. The distance unit in the figure is angstrom (Å)

One of the challenges of a single-trajectory-based MD analysis for orientation during deposition is that the time scale needed to observe the entirety of lateral degrees of freedom for a single molecule is much longer than that of a typical MD simulation. As such, we introduced 50 independent deposition events per dopant, instead of relying upon a single MD trajectory for each. 50 target dopant molecules were distributed in the vacuum slab of the periodic substrate model at un-overlapped locations with different orientations for the deposition simulation. Each target molecule

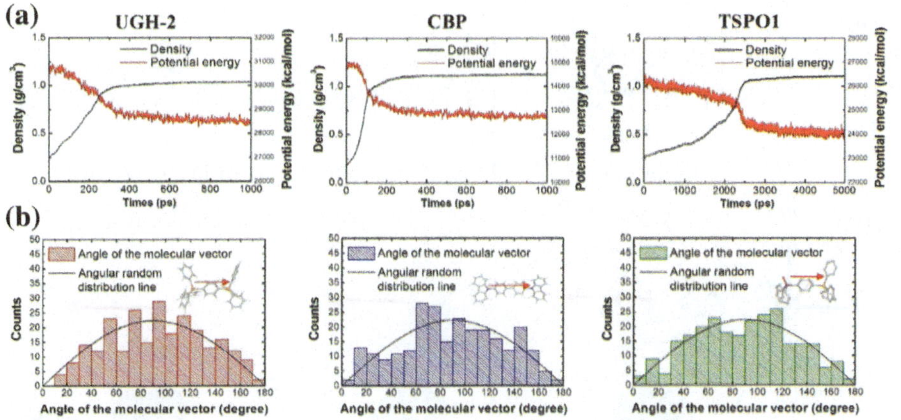

Fig. 3.7 Potential energy, volume, and molecular distribution of the organic substrates prepared by MD simulation. **a** Change of the density and total potential energy of the substrates consisting of 256-number of UGH-2, CBP, and TSPO1 molecules, respectively, in the simulation of packing the molecules at 300 K and 1 atm. **b** Angular distributions of the 256-number of molecular vectors (red arrows) in the substrates. Their distribution followed the random angular distribution line, indicating the amorphous substrates were formed by the MD simulation

was individually dropped onto the substrate under vacuum at 300 K. Translational motion of the host molecules at the bottom of the substrate was restrained in order to avoid the drift of the system. The deposition simulation used an NVT ensemble for a duration of 6000 ps with a time step of 2 fs and configurations of the system were recorded every 6 ps. Finally, EDOs of the phosphors and the molecular angles (φ_C) were analyzed using Eq. (3.1) from the configurations. The analysis is based upon an assumption that the characteristic time to determine the orientation of dopants is in same scale of which the intermolecular interaction converges after the deposition of a dopant.

Materials

Chemical structures of the materials used in this study are depicted in Fig. 3.8a and b. Three heteroleptic iridium complexes of $Ir(ppy)_2tmd$, $Ir(3',5',4\text{-}mppy)_2tmd$ [22], and $Ir(dmppy\text{-}ph)_2tmd$ [20] possessing high Θ values were adopted to investigate the effect of the phosphor molecular structure. The molecular C_2 symmetry axis toward the center of the ancillary ligand from the origin located at the Ir atom was set as $\boldsymbol{m}_z$, the orthogonal vector to $\boldsymbol{m}_z$ normal to the molecular Ir–O–O plane was set as $\boldsymbol{m}_x$, and $\boldsymbol{m}_y$ was determined by a cross product of $\boldsymbol{m}_z$ and $\boldsymbol{m}_x$ in the dopants. The triplet TDM vectors of the three Ir-complexes align along the direction of the iridium atom to the pyridine rings by ^{3}MLCT as displayed in Fig. 3.7a. Coordinates of the TDM vectors of $Ir(ppy)_2tmd$, $Ir(3',5',4\text{-}mppy)_2tmd$, and $Ir(dmppy\text{-}ph)_2tmd$ were [$\varphi_M = 88°, \theta = 147°$], [$\varphi_M = 89°, \theta = 141°$], and [$\varphi_M = 89°, \theta = 156°$], respectively, indicating that the substituents at the 4-position of the pyridine of the main ligands do not change the direction of triplet TDM vectors much.

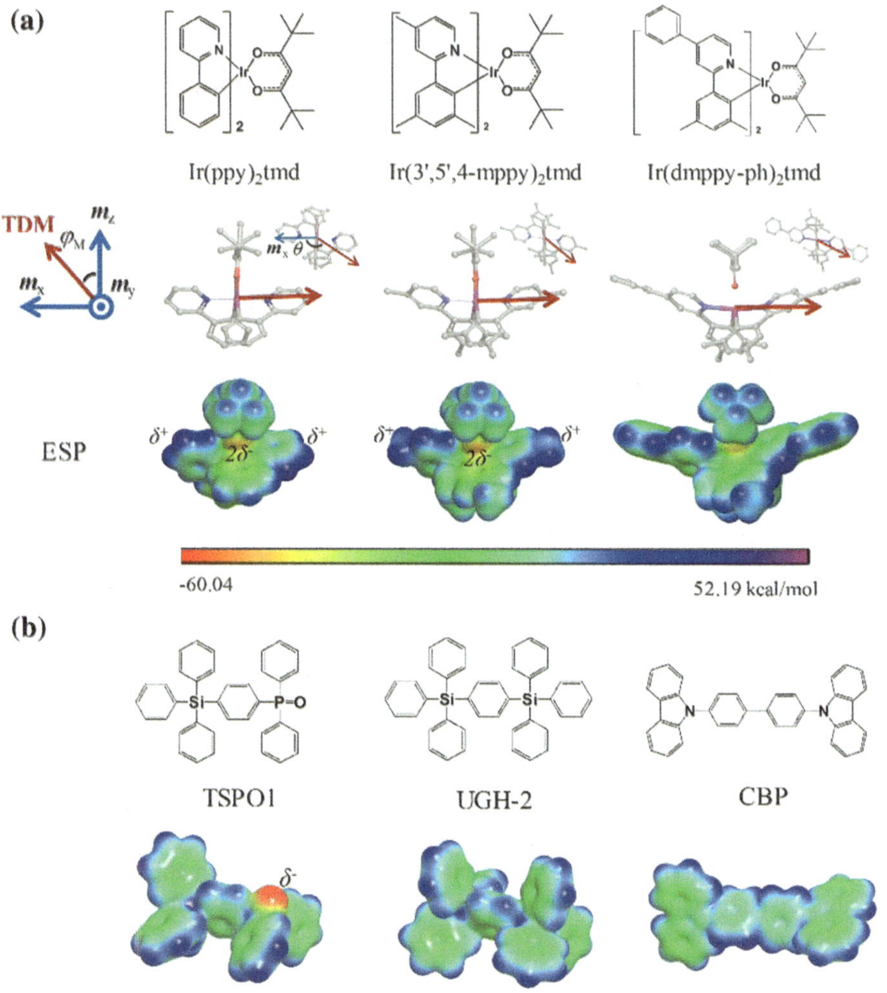

Fig. 3.8 Iridium complexes and host materials. **a** Chemical structures, transition dipole moment vectors, and electrostatic potentials of $Ir(ppy)_2tmd$, $Ir(3',5',4\text{-}mppy)_2tmd$, and $Ir(dmppy\text{-}ph)_2tmd$ phosphors. There are linear quadrupoles in the ground state of $Ir(ppy)_2tmd$ and $Ir(3',5',4\text{-}mppy)_2tmd$ with quadrupole moments along the principal axes of $Q_{xx,yy,zz} = [25.2, -13.0, -12.2]$ and $[27.1, -12.8, -14.3]$ Debye $\cdot$ Å^2, respectively. **b** Chemical structures and electrostatic potentials of UGH-2, CBP, and TSPO1 host molecules. The electrostatic potentials are projected on the iso-surface of electron density of 0.005 electrons/bohr^3. Optimization of the molecular structures were demonstrated using B3LYP method and LACVP** basis set for the phosphors and 6–31 g(d)** for the host materials, respectively. SOC-TDDFT of the phosphors were carried out using B3LYP method and DYALL-2ZCVP_ZORA-J-PT-GEN basis set

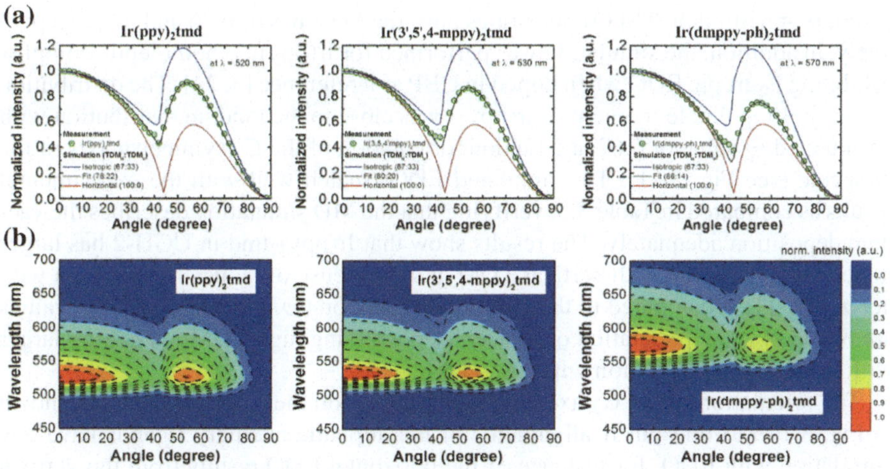

Fig. 3.9 Angle-dependent PL analysis of Ir(ppy)$_2$tmd, Ir(3′,5′,4-mppy), and Ir(dmppy-ph)$_2$tmd doped in 30 nm of TSPO1 layers. **a** Analysis of the EDO at peak wavelength of each emitter (520, 530, and 570 nm). Measured emission patterns (scatters) are located between the lines with an isotropic orientation (blue) and a fully horizontal orientation (red). The best fitted fractions of the horizontal to vertical emitting dipole moment were 78:22, 79:21, and 85:15, for Ir(ppy)$_2$tmd, Ir(3′,5′,4-mppy), and Ir(dmppy-ph)$_2$tmd, respectively. **b** Measured (surfaces) and calculated angular emission spectra (broken lines) having the orientation that have been determined by the analyses at the peak wavelengths

Diphenyl-4-triphenylsilyphenyl-phosphineoxide (TSPO1), 1,4-bis (triphenylsilyl)benzene (UGH-2), and 4,4′-bis(*N*-carbazolyl)-1,1′-biphenyl (CBP) were selected as host materials to investigate the effect of ground state dipole and conjugation length of host on the EDO. TSPO1 has large permanent dipole moment due to the polar phosphine oxide group and the asymmetric structure, while UGH-2 and CBP molecules have small ground state dipole moments compared to TSPO1 due to the symmetric structures and the less polar groups. On the other hand, CBP has longer conjugation length than UGH-2 and TSPO1, indicating that CBP has larger polarizability than UGH-2.

Experimentally, Ir(ppy)$_2$tmd exhibited the Θ values of 0.60, 0.75 and 0.78 when doped in the UGH-2, CBP, and TSPO1 layers, respectively. Ir(3′,5′,4-mppy)$_2$tmd and Ir(dmppy-ph)$_2$tmd doped in TSPO1 layers have enhanced horizontal dipole orientation with the Θ values of 0.80 and 0.86, respectively (Fig. 3.9).

Simulation results

Figure 3.10a exhibits the histograms of the EDO resulting from the deposition simulation. The blue lines represent the probability density of $TDM_H(\sin^2 \varphi_L)$ of an arbitrary vector. The green lines exhibit the deviations of the population from the random distribution. The simulated Θ values of Ir(ppy)$_2$tmd were 0.63, 0.72 and 0.74 on the UGH-2, CBP and TSPO1 substrates, respectively. Ir(3′,5′,4-mppy)$_2$tmd and

$Ir(dmppy\text{-}ph)_2tmd$ on TSPO1 substrates have the Θ values of 0.76 and 0.82, respectively. In addition, the simulation was performed for $Ir(ppy)_3$, a homoleptic complex exhibiting isotropic EDO when doped in CBP as a reference [3, 24]. The distribution of the emitting dipole moment of $Ir(ppy)_3$ was close to the random distribution with a simulated Θ value of 0.67 and random orientation of the C_3 symmetry axis of the molecule (see Fig. 3.11). The simulated EDOs match well with the experimental results as compared in Table 3.3, verifying that the MD simulation describes the vacuum deposition adequately. The results show that $Ir(ppy)_2tmd$ in UGH-2 has larger molecular population with vertical TDM at the expense of reduced population with horizontal TDM compared to the random distribution ($\Theta = 0.67$). Higher Θ values are obtained when population of molecules possessing high TDM_H is getting larger with the reduced population with low TDM_H.

The orientation of the C_2 axes of the phosphors on the organic layers are shown in Fig. 3.10b to find out if alignments of the aliphatic ancillary ligands have any correlation with EDO, for instance, if the horizontal EDO results from the vertical alignment of ancillary ligands with respect to the substrate [21, 25]. One expects that the distribution function follows $\sin \varphi_c$ (blue line) if the orientation is random. We can extract several interesting results from Fig. 2.10b: (1) The orientation of the ancillary ligand of the Ir-complexes has broad distributions for all the deposited films. (2) Host effect on EDO is independent of alignments of the ancillary ligand. The total distribution of the C_2 axis of $Ir(ppy)_2tmd$ are similar on the UGH-2, CBP, and TSPO1 layers with the average φ_C of 70°, but the EDO on UGH-2 host is different from the EDOs on other two hosts. (3) The orientation of the C_2 axes of $Ir(3,5',4\text{-}mppy)_2tmd$ and $Ir(dmppy\text{-}ph)_2tmd$ on TSPO1 are more random (closer to $\sin \varphi_c$) even though they possess higher Θ values than $Ir(ppy)_2tmd$. The random distributions are observed even in the region with high horizontal alignment of the emitting dipole moment (green regions in the stacked histogram with $0.95 \leq TDM_H \leq 1$). (4) The dopant molecules with vertical TDM (red regions in the stacked histogram with $0 \leq TDM_H \leq 0.3$) have φ_C close to 90° for all the system, indicating that the ancillary ligands align parallel to the surface. All the results show that there is little correlation between the orientation of TDMs and alignment of the ancillary ligands.

3.2.3 Discussion

The size of the substrates turns out to be large enough to simulate the vacuum deposition of the phosphorescent dyes adequately, as confirmed by the similar results obtained on a larger substrate consisting of 1024 molecules. We demonstrated the deposition simulation with larger number of host molecules. Figure 3.12a exhibits the equilibrated 1024-molecule-TSPO1 substrate. Figure 3.12b and c show histograms of TDM_H and angle of the C_2 axis of $Ir(ppy)_2tmd$, respectively, deposited on the 1024-molecule substrate. The simulated EDO value (Θ, ratio of horizontal transition dipole moment) was 0.74 which corresponds to the value of the deposition using the substrate consisting of 256 TSPO1 molecules. The distribution of the C_2 axis vector

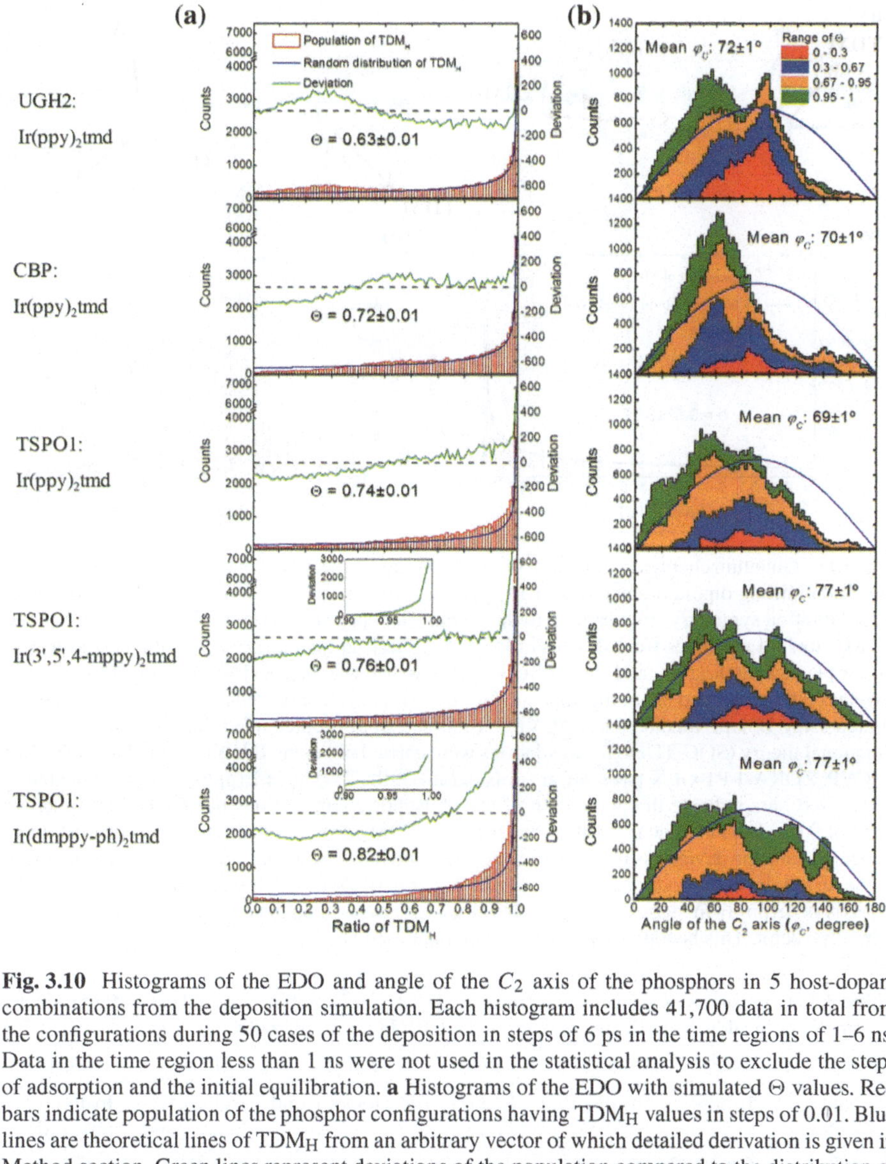

Fig. 3.10 Histograms of the EDO and angle of the C_2 axis of the phosphors in 5 host-dopant combinations from the deposition simulation. Each histogram includes 41,700 data in total from the configurations during 50 cases of the deposition in steps of 6 ps in the time regions of 1–6 ns. Data in the time region less than 1 ns were not used in the statistical analysis to exclude the steps of adsorption and the initial equilibration. **a** Histograms of the EDO with simulated Θ values. Red bars indicate population of the phosphor configurations having TDM_H values in steps of 0.01. Blue lines are theoretical lines of TDM_H from an arbitrary vector of which detailed derivation is given in Method section. Green lines represent deviations of the population compared to the distribution of TDM_H of an arbitrary vector. (Inset: enlarged deviation in the region of $0.8 \leq TDM_H \leq 1$) **b** Stacked histogram of the angle of the C_2 axis of phosphors and mean angles. Populations of the vector are plotted in steps of 2°. Distribution of the angle in different ranges of TDM_H is distinguished by different colors

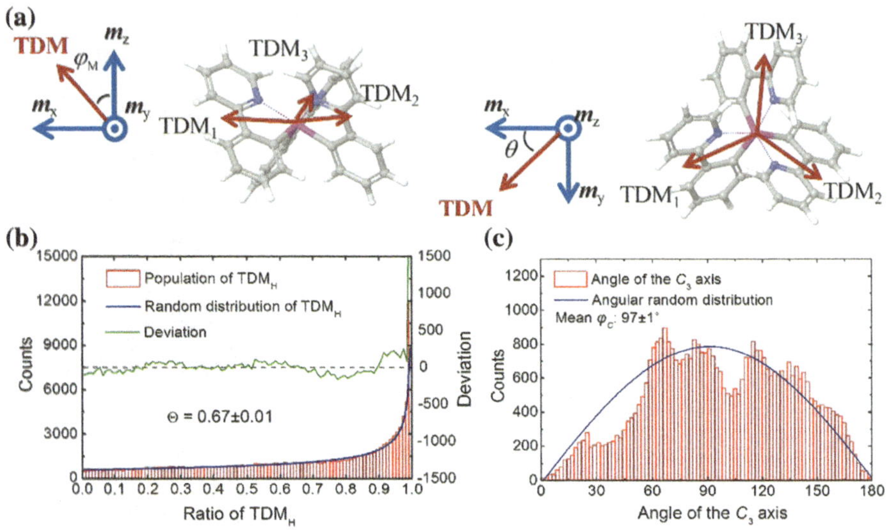

Fig. 3.11 Quantum chemical simulation of the TDMs and molecular dynamic simulation of molecular and emitting dipole orientations of Ir(ppy)$_3$. **a** Three triplet TDM vectors of Ir(ppy)$_3$ with a 3-fold rotation symmetry from iridium to three equivalent ppy ligands by ^{3}MLCT. The C_3 symmetry axis toward pyridines from the origin located at the Ir atom was set as m_z, the vector normal to the plane including m_z and one of Ir-N vector was set as m_y, and m_x was determined by a cross product of m_y and m_z in the dopants. Optimization of the molecular structures were demonstrated using B3LYP method and LACVP** basis set. Spin-orbit coupled time-dependent density functional theory (SOC-TDDFT) calculations were carried out using B3LYP method and DYALL-2ZCVP_ZORA-J-PT-GEN basis set. **b** A histogram of the TDM$_H$ of Ir(ppy)$_3$ with a simulated Θ value. Red bars indicate the population of the phosphor configurations having TDM$_H$ values in steps of 0.01. The blue line is the theoretical line of TDM$_H$ from an arbitrary vector and the green line represents the deviations between red bars and blue lines. Note that 125100 data were included in the histogram by a product of 41,700 frames and three TDMs. **c** A histogram of the angle of the C_3 axis of Ir(ppy)$_3$ in steps of 2°. The blue line represents an angular random distribution of an arbitrary vector. This histogram includes 41,700 data in total

Table 3.3 Comparison of simulated and measured EDOs in 5 combinations of host and heteroleptic Ir complexes in addition to Ir(ppy)$_3$ doped in the CBP layer for reference

Host	CBP	UGH-2	CBP	TSPO1	TSPO1	TSPO1
Dopant	Ir(ppy)$_3$	Ir(ppy)$_2$tmd	Ir(ppy)$_2$tmd	Ir(ppy)$_2$tmd	Ir(3′,5′,4-mppy)$_2$tmd	Ir(dmppy-ph)$_2$tmd
Simulation	0.67:0.33	0.63:0.37	0.72:0.28	0.73:0.27	0.76:0.24	0.82:0.18
Measurement	0.67:0.33	0.60:0.40	0.75:0.25	0.78:0.22	0.80:0.20	0.86:0.14

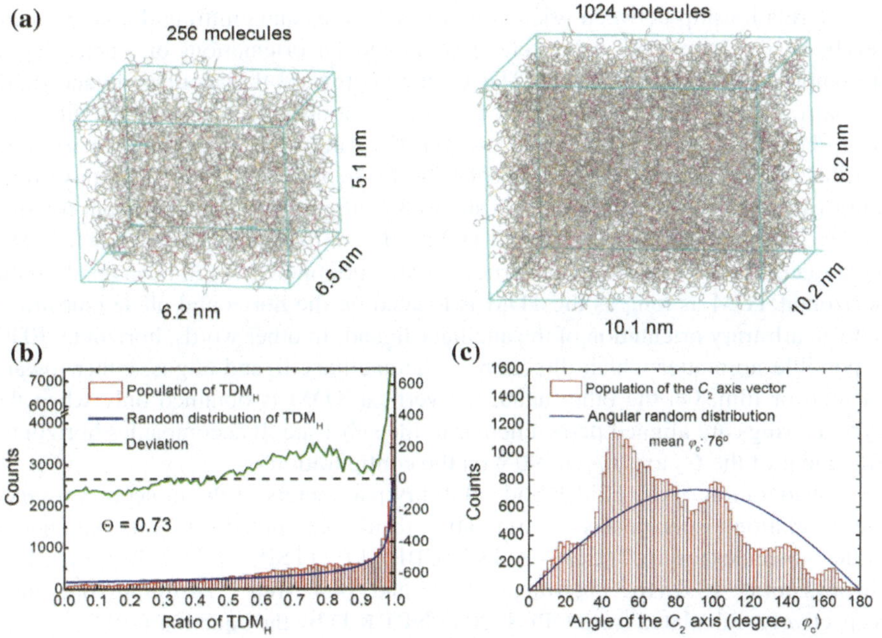

Fig. 3.12 **a** Comparison of TSPO1 substrates consisting of 256 and 1024 molecules. **b** Histograms of TDM_H and **c** φ_C demonstrated by the deposition simulation of Ir(ppy)$_2$tmd on the 1024-molecule TSPO1 substrate

was broad with mean angle of 74°, exhibiting no large difference of the molecular distribution with the simulation results using the 256-molecule substrate.

We performed the MD simulation by depositing each target molecule on an organic substrate at a time and obtained statistics by repeating for 50 molecules deposited on different positions of the organic substrates so that the aggregation effect of the Ir complexes is neglected. Very good consistency of the simulated EDO and experimental values clearly indicates that aggregation is not a necessary condition for the alignment of heteroleptic Ir complexes.

Alignment of aliphatic ligands of heteroleptic Ir complexes to vacuum (model 3) is not required for preferred horizontal EDO either as shown in Fig. 3.10. Much larger portion of the aliphatic ligand (–tmd group) of Ir(ppy)$_2$tmd molecules align to the vacuum side ($0° < \varphi_C < 90°$ in Fig. 3.10b) than Ir(3′,5′,4-mppy)$_2$tmd and Ir(dmppy-ph)$_2$tmd molecules. However, the Θ value of Ir(ppy)$_2$tmd is much lower than Ir(3′,5′,4-mppy)$_2$tmd and Ir(dmppy-ph)$_2$tmd. These results are the reverse direction from the prediction based on the model and clearly demonstrate, therefore, that alignment of aliphatic ligands to the vacuum side is not a necessary condition for the alignment of EDO in heteroleptic Ir complexes. The reason why it is not required can understood from the following consideration.

The relationship between orientation of molecules and emitting dipole moment can be easily figured out using schematic molecular orientations of a heteroleptic Ir complex shown in Fig. 3.10a. The C_2 axis is toward the ancillary ligand (dark blue arrows) and the TDM vector (red arrows) is approximately along the direction from the iridium center to one of the pyridine rings. The alignment of iridium-pyridines determines the orientations of TDM for Ir(ppy)$_2$tmd, Ir(3′,5′,4-mppy)$_2$tmd, and Ir(dmppy-ph)$_2$tmd. Figure 3.10a shows 5 configurations with different rotation angles of the C_2 axis for the horizontal TDM and 1 configuration for the vertical TDM. Rotation of the C_2 axis from the vertical to the horizontal direction can result in the horizontal TDM as long as the TDM is located on the horizontal plane (substrate) with an arbitrary orientation of the ancillary ligand. In other words, horizontal EDO is possible no matter which direction of the ancillary ligand aligns, either toward vacuum or film. On the other hand, the vertical TDM is obtained only when the pyridine rings are aligned perpendicular to the substrate. It accompanies horizontal alignment of the C_2 axis ($\varphi_C \sim 90°$) on the configuration.

Consideration of the Hildebrand solubility parameters of the molecules used in the simulation also supports the simulation results. Predicted solubility parameters of the molecules are 16.2 (UGH-2), 18.5 (CBP), 17.2 (TSPO1), 15.3 (Ir(ppy)$_3$), 14.7 (Ir(ppy)$_2$tmd), 13.8 (Ir(3′,5′,4-mppy)$_2$tmd) and 14.4 MPa$^{1/2}$ (Ir(dmppy-ph)$_2$tmd), respectively, calculated from OPLS_2005 NPT MD by the equation [29]:

$$\delta = \left(\frac{\Delta E_{\mathrm{v}}}{\mathrm{V_m}}\right)^{1/2} \tag{3.3}$$

where ΔE_{v} is the internal energy change of vaporization, and $\mathrm{V_m}$ is the molar volume, respectively. In general, the differences in the solubility parameters ($\Delta\delta$) between two components in a chemical mixture can be an indicator of the degree of miscibility, with smaller and larger values of $\Delta\delta$ indicating more and less miscible, respectively. In this work, the host and the phosphor $\Delta\delta$ are much less than 7 MPa$^{1/2}$, suggesting that all the phosphors are miscible with the hosts [30]. However, $\Delta\delta$'s between Ir(ppy)$_3$ and the hosts are smaller than between Ir(ppy)$_2$tmd and the hosts. Since the difference comes from the ppy and tmd groups, it follows that the tmd group is less miscible in the host substrates than the main ligand ppy group, which explains the orientation of the aliphatic ancillary ligand toward vacuum side for Ir(ppy)$_2$tmd. On the other hand, the difference in the solubility parameters among Ir(ppy)$_2$tmd, Ir(3′,5′,4-mppy)$_2$tmd and Ir(dmppy-ph)$_2$tmd comes from the difference in main ligands. The reduced solubility of Ir(3′,5′,4-mppy)$_2$tmd and Ir(dmppy-ph)$_2$tmd indicates that both 3′,5′,4-mppy and dmppy-ph groups are less miscible to the host than ppy of Ir(ppy)$_2$tmd and less preference to attachment to the substrate compared to ppy group. Therefore, the orientations of the ancillary ligand of the two phosphors are more randomized during the deposition, consistent with the simulated distributions in Fig. 3.10b.

Non-bonded interaction energy was calculated from the MD simulation to investigate if the intermolecular interaction between the phosphor and neighbor host molecules is responsible for the spontaneous molecular alignments of the phosphors on the surfaces. Figure 3.13b depicts the correlation between non-bonded interaction energy and orientation of the emitting dipole moment of the phosphors in the five different host-dopant systems. There is a broad energy trap of ~3 kcal/mol in the region of $TDM_H = 0.1–0.5$ for Ir(ppy)$_2$tmd on the UGH-2 host and the energy increases with further increasing of TDM_H, thereby resulting in rather vertical EDO compared to random orientation because the population of TDM_H is expected to be concentrated in the regions of low (large) non-bonded interaction energy. On the other hand, non-bonded interaction energies of Ir(ppy)$_2$tmd on CBP and TSPO1 layers and the energies of Ir(3′,5′,4-mppy)$_2$tmd and Ir(dmppy-ph)$_2$tmd on TSPO1 are lowered as TDM_H increases. As a result, molecular alignment with horizontal TDM is energetically preferred when they are deposited onto the organic semiconducting layers. Furthermore, much lower energies were obtained from Ir(3′,5′,4-mppy)$_2$tmd and Ir(dmppy-ph)$_2$tmd than Ir(ppy)$_2$tmd on the TSPO1 layer, indicating that the increased EDOs are also related to the stabilization by neighbor molecules. The calculated non-bonded interaction energy and the statistical results indicate that the host-dopant interaction plays a pivotal role for orienting heteroleptic Ir complexes and the force applies to the alignment of the iridium-pyridine bonds of the phosphors toward horizontal direction.

The type and magnitude of the non-bonded interactions are different for different phosphors and hosts, leading to different EDOs as shown in Fig. 3.14 Dispersion interaction between the aromatic ligands and nearest host molecules contributes to the molecular alignment of the phosphors and electrostatic interaction between them helps the further alignments. For instance, Ir(ppy)$_2$tmd and Ir(3′,5′,4-mppy)$_2$tmd have a quadrupole composed of two dipoles from pyridines (δ + charge) to Ir atom ($2\delta-$ charge). If there is a dipole in host molecule (i.e., TSPO1), dipole and quadrupole interaction [$-P = O^{\delta-}$ and $^{\delta+}$H(pyridine)] anchors one phosphor molecule to two host molecules, leading to rather horizontal orientation of iridium-pyridines bond of the phosphors which is approximately parallel to the TDM. In contrast, if host molecule has positive surface potential [i.e., $^{\delta+}$(phenyl)$_3$-Si-phenyl-Si-(Phenyl)$_3^{\delta+}$ in UGH-2], there must be repulsive force between pyridine of phosphors and host molecules so that pyridine ring must be pushed to vacuum. Dispersion force between the conjugated phenyl substituents of Ir(dmppy-ph)$_2$tmd and nearest neighbors anchors the pyridines onto the surface as well and lowers the energy with the molecular long axis lying on the surface. Meanwhile, random EDO of Ir(ppy)$_3$ is attributed to three intermolecular interaction sites, resulting in random orientation of the molecule.

Figure 3.15a, b, and c exhibit the representative molecular behaviors of phosphors and nearest host molecules during the deposition out of 50 cases with different host-dopant combinations of UGH-2:Ir(ppy)$_2$tmd, TSPO1:Ir(ppy)$_2$tmd, and TSPO1:Ir(dmppy-ph)$_2$tmd, respectively. Large vibrations, rotations and diffusions of Ir(ppy)$_2$tmd on the surface of UGH-2 layer without lowering the energy were observed in the trajectory shown in Fig. 3.15a. The perpendicular alignment of pyridines occasionally formed on the surface resulted in vertical emitting dipole moment in average. On the other hand, a hydrogen atom at one of pyridines of Ir(ppy)$_2$tmd faced toward an oxygen atom of TSPO1 with the $-P = O \cdot$ H(pyridine) distance around 0.4 nm at $t = 2472$ ps and $t = 4560$ ps, thereby the parallel align-

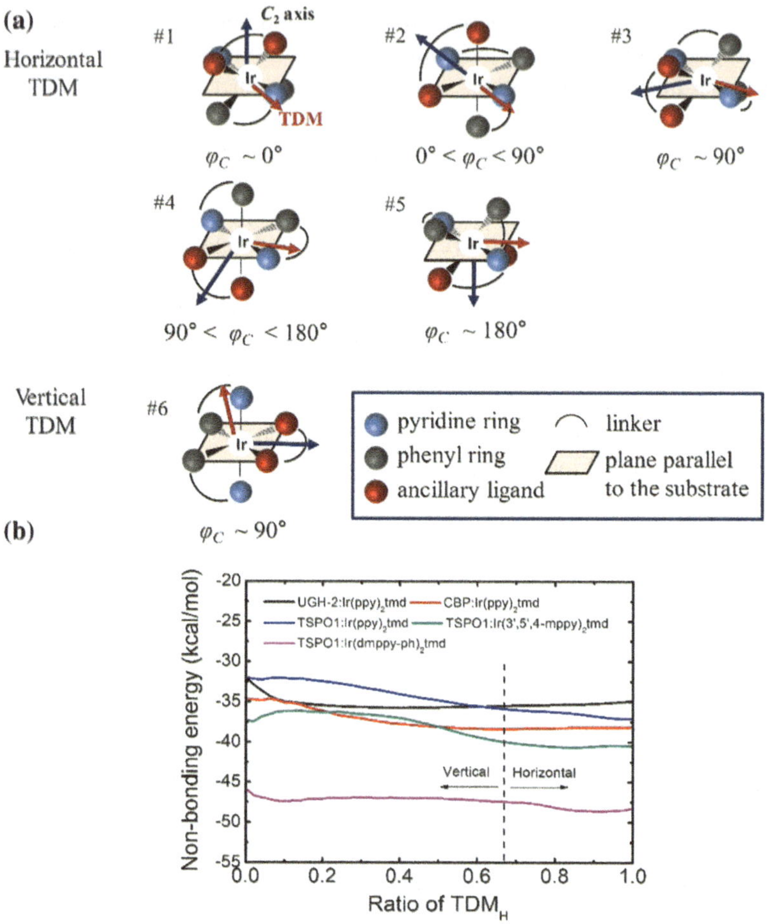

Fig. 3.13 Schematic illustration of heteroleptic Ir-complexes having horizontal and vertical transition dipole moments and non-bonded interaction energy of phosphors depending on the dipole orientation. **a** Blue, gray, and red spheres at the octahedral sites represent pyridine rings, phenyl rings, and the ancillary ligand (–tmd), respectively. Dark blue and red arrows indicate the molecular C_2 axis and the TDM vector, respectively. Five configurations of the molecule for horizontal TDM and one configuration for vertical TDM are illustrated depending on the angle of the C_2 axis. **b** Calculated bon-bonding energy with cut-off radius of 0.9 nm of each atom of the phosphors as a function of TDM_H in the five host-dopant systems

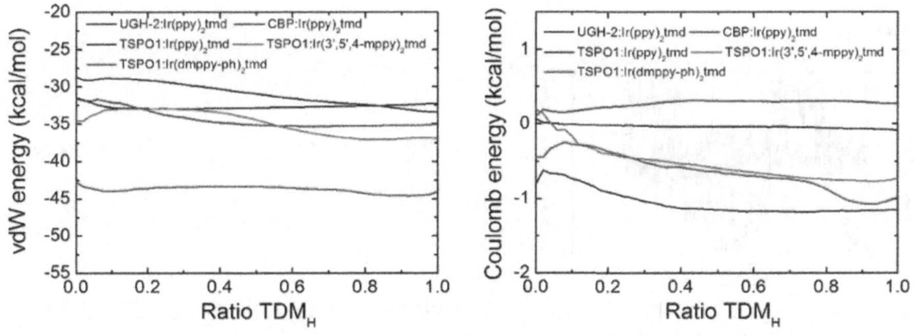

Fig. 3.14 Van der Waals and Coulomb interaction energies as a function of the emitting dipole orientation

ment of the Ir-pyridines of $Ir(ppy)_2tmd$ to the surface. Larger quadrupole moment of $Ir(3',5',4\text{-}mppy)_2tmd$ than that of $Ir(ppy)_2tmd$ increased the strength of quadrupole-dipole interaction and resulted in enhancement of fraction of the horizontal dipole compared to $Ir(ppy)_2tmd$. The horizontal EDO of $Ir(ppy)_2tmd$ in CBP could be understood by the dipole inducement in carbazole groups of CBP when the positive pole of pyridine approaches but the interaction strength for the iridium-pyridines alignment between $Ir(ppy)_2tmd$-CBP is smaller than that between $Ir(ppy)_2tmd$-TSPO1. Compared to the former cases, the picture of $Ir(dmppy\text{-}ph)_2tmd$ shown in Fig. 3.15c is rather simple. The phosphor deposited onto the TSPO1 layer was stabilized after short time of deposition to the one with the horizontal iridium-pyridine-phenyl alignment and maintain the configuration. Much lower (larger) non-bonded interaction energy of $Ir(dmppy\text{-}ph)_2tmd$ restrained rotation of the molecule and the molecular configurations are easily fixed on the surface, resulting in much enhancement of horizontal EDO was achieved by the substitutions.

3.2.4 Conclusion

We investigated the origin of the molecular orientation and EDO of doped heteroleptic iridium complexes in vacuum-deposited organic layers using MD simulations and QM analyses in direct comparison with experimental observation. Careful analyses on the simulation results revealed that molecular alignments of the phosphors are spontaneous by local electrostatic and van der Waals interaction with nearest host molecules interacting in a smaller scale than a molecule. Orientation of the TDM vector of the phosphors on the organic surfaces follows the direction of the ligand mainly participating in optical transition, such as pyridines in ppy, in the molecular alignment, whereas the alignment of ancillary ligand does not have a direct correlation with the EDO. Attractive interactions between pyridines of a phosphor and CBP (quadrupole–induced dipole interaction) or TSPO1 (quadrupole–dipole inter-

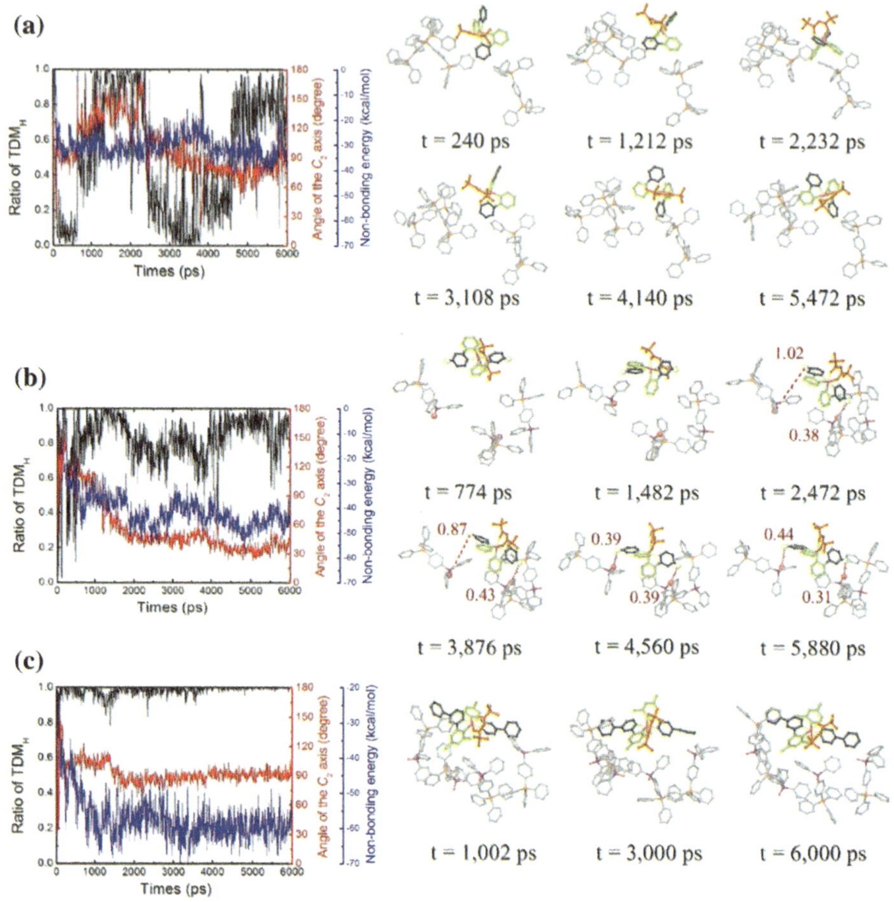

Fig. 3.15 Representative configurations and non-bonded interaction energy of the phosphor on the surface during the deposition. Snapshots of local configurations and time-dependent trajectories of the EDO, angle of the C_2 axis, and non-bonded interaction energy up to 6 ns are depicted together. The ancillary ligand and pyridine rings of the phosphors at the octahedral sites are colored by red and blue, respectively. **a** Ir(ppy)$_2$tmd deposited onto the UGH-2 layer has continuous rotation and the occasionally observed perpendicular alignment of pyridines with respect to the substrate results in vertical EDO. **b** Ir(ppy)$_2$tmd anchors on the surface of TSPO1 layer by local quadrupole-dipole interaction between the two nearest host molecules located at both sides. The hydrogen atoms at both pyridines of Ir(ppy)$_2$tmd and the oxygen atoms of TSPO1 connected by a broken line were the plausible binding sites. The distances between the two atoms (broken lines) are getting closer until around 0.4 nm as the time increases and a host-dopant-host pseudo-complex is formed with the parallel alignment of pyridines with respect to the substrate. **c** Ir(dmppy-ph)$_2$tmd deposited onto the TSPO1 layer are less mobile than Ir(ppy)$_2$tmd with the low non-bonded interaction energy by the configuration of large dispersion force energy along the direction of TDM

action) anchor the phosphor onto host molecules with the parallel iridium-pyridine alignment, thereby increasing the horizontal EDO. Ir(3′,5′,4-mppy)$_2$tmd has larger quadrupole moment than Ir(ppy)$_2$tmd resulted in further molecular alignment for the horizontal emitting dipole moment. The increase of the dispersion force along the direction of TDM was also effective on control of the molecular orientation for the horizontal EDO with lowered non-bonded interaction energy.

3.2.5 Methods

Quantum mechanical calculations and molecular dynamics simulations

Density functional theory (DFT) was used to obtain molecular geometries and electrostatic potentials of host and phosphors. Triplet TDMs of the same phosphors from T_1 to S_0 were also calculated via SOC-TDDFT. The DFT and SOC-TDDFT calculations and the follow-up analyses were performed with Schrödinger Materials Science Suite [26] along with the quantum chemical engine, Jaguar [31]. A TDM having the largest oscillator strength among the three degenerated states of T_1 level (T_x, T_y, and T_z) obtained from the density functional calculations was used in this study as a representative transition dipole moment. All MD simulations were performed by Desmond [32, 33], a molecular dynamics engine implemented in the Schrodinger Materials Science Suite. Equilibration simulations prior to deposition were performed in NPT ensembles where pressure and temperature were set constant via Nose-Hoover chain and Martyna-Tobias-Klein method, respectively. There were no explicit constraints to geometry and/or positions to any of the molecules that were introduced in the simulation box. The simulations were performed over NVIDIA general-purpose GPU cards (K80).

Rotation matrix method

Rotation matrix method was used for transformation of the TDM vector from the molecular coordinate to the laboratory coordinate. Rotation angles of α, β, and γ are defined as the clockwise rotations to laboratories axes of $\boldsymbol{n}_x$, $\boldsymbol{n}_y$, and $\boldsymbol{n}_z$ axes, respectively. Then, the rotation matrixes for α, β, and γ rotations are followings:

$$R_\alpha = \begin{pmatrix} 1 & 0 & 0 \\ 0 & \cos\alpha & \sin\alpha \\ 0 & -\sin\alpha & \cos\alpha \end{pmatrix}, \tag{3.4}$$

$$R_\beta = \begin{pmatrix} \cos\beta & 0 & -\sin\beta \\ 0 & 1 & 0 \\ \sin\beta & 0 & \cos\beta \end{pmatrix}, \tag{3.5}$$

$$R_\gamma = \begin{pmatrix} \cos\gamma & \sin\gamma & 0 \\ -\sin\gamma & \cos\gamma & 0 \\ 0 & 0 & 1 \end{pmatrix}. \tag{3.6}$$

Sequential $\alpha\beta\gamma$ rotations of the dopant molecules were extracted in every configurations of the MD simulation. A product of the three rotation matrixes gives a matrix representing the orientation of the dopant molecule by

$$R_{\text{total}} = R_\gamma R_\beta R_\alpha. \tag{3.7}$$

Finally, the TDM vectors in the laboratory coordinate were obtained by

$$\text{TDM}_{\text{Lab}} = R_{\text{total}} \text{TDM}_{\text{Mol}}. \tag{3.8}$$

Calculation of a probability density function of TDM$_H$.

Integration of a probability density function, f, indicates a probability of a variable X between $X = a$ and b.

$$P(a < X < b) = \int_a^b f_X(x)dx, \tag{3.9}$$

where

$$\int_{-\infty}^{\infty} f_X(x)dx = 1. \tag{3.10}$$

To calculate the probability density function of $\sin^2\varphi$ from an arbitrary vector, we define an arcsine function of

$$y = \arcsin(\sqrt{x}),\ 0 \le x \le 1, \tag{3.11}$$

which is a reversed function of $y = \sin^2 x$. For a monotonic function, the variables are related by

$$f_X(x) = f_Y(y)\frac{dy}{dx}. \tag{3.12}$$

If we put $f_Y(y) = \sin(y)$ and $\frac{dy}{dx} = \frac{1}{2\sqrt{x-x^2}}$ into Eq. (3.12), the probability density function (f_X) is obtained as

$$f_X(x) = \frac{1}{2\sqrt{1-x}}. \tag{3.13}$$

References

1. Yokoyama D (2011) Molecular orientation in small-molecule organic light-emitting diodes. J Mater Chem 21:19187. https://doi.org/10.1039/c1jm13417e
2. Flämmich M et al (2011) Oriented phosphorescent emitters boost OLED efficiency. Org Electron 12:1663–1668
3. Liehm P et al (2012) Comparing the emissive dipole orientation of two similar phosphorescent green emitter molecules in highly efficient organic light-emitting diodes. Appl Phys Lett 101:253304
4. Kim S-Y et al (2013) Organic light-emitting diodes with 30% external quantum efficiency based on a horizontally oriented emitter. Adv Func Mater 23:3896–3900. https://doi.org/10.1002/adfm.201300104
5. Kim KH et al (2014) Phosphorescent dye-based supramolecules for high-efficiency organic light-emitting diodes. Nat Commun 5:4769. https://doi.org/10.1038/ncomms5769
6. Wasey JAE, Barnes WL (2000) Efficiency of spontaneous emission from planar microcavities. J Mod Opt 47:725–741. https://doi.org/10.1080/09500340008233393
7. Schmidt TD et al (2011) Evidence for non-isotropic emitter orientation in a red phosphorescent organic light-emitting diode and its implications for determining the emitter's radiative quantum efficiency. Appl Phys Lett 99:163302. https://doi.org/10.1063/1.3653475
8. Kim KH, Moon CK, Lee JH, Kim SY, Kim JJ (2014) Highly efficient organic light-emitting diodes with phosphorescent emitters having high quantum yield and horizontal orientation of transition dipole moments. Adv Mater 26:3844–3847. https://doi.org/10.1002/adma.201305733
9. Flämmich M et al (2010) Orientation of emissive dipoles in OLEDs: quantitative in situ analysis. Org Electron 11:1039–1046
10. Penninck L, Steinbacher F, Krause R, Neyts K (2012) Determining emissive dipole orientation in organic light emitting devices by decay time measurement. Org Electron 13:3079–3084
11. Shin H et al (2014) Blue phosphorescent organic light-emitting diodes using an exciplex forming co-host with the external quantum efficiency of theoretical limit. Adv Mater 26:4730–4734. https://doi.org/10.1002/adma.201400955
12. Taneda M, Yasuda T, Adachi C (2011) Horizontal orientation of a linear-shaped platinum(II) complex in organic light-emitting diodes with a high light out-coupling efficiency. Appl Phys Express 4:071602. https://doi.org/10.1143/apex.4.071602
13. Mayr C, Taneda M, Adachi C, Brütting W (2014) Different orientation of the transition dipole moments of two similar Pt(II) complexes and their potential for high efficiency organic light-emitting diodes. Org Electron 15:3031–3037. https://doi.org/10.1016/j.orgel.2014.07.042
14. Komino T, Tanaka H, Adachi C (2014) Selectively controlled orientational order in linear-shaped thermally activated delayed fluorescent dopants. Chem Mater 26:3665–3671. https://doi.org/10.1021/cm500802p
15. Komino T, Nomura H, Koyanagi T, Adachi C (2013) Suppression of efficiency roll-off characteristics in thermally activated delayed fluorescence based organic light-emitting diodes using randomly oriented host molecules. Chem Mater 25:3038–3047. https://doi.org/10.1021/cm4011597
16. Lamansky S et al (2001) Highly phosphorescent bis-cyclometalated iridium complexes: synthesis, photophysical characterization, and use in organic light emitting diodes. J Am Chem Soc 123:4304–4312
17. Frischeisen J, Yokoyama D, Adachi C, Brütting W (2010) Determination of molecular dipole orientation in doped fluorescent organic thin films by photoluminescence measurements. Appl Phys Lett 96:29
18. Moon CK, Kim SY, Lee JH, Kim JJ (2015) Luminescence from oriented emitting dipoles in a birefringent medium. Opt Express 23:A279–A291. https://doi.org/10.1364/OE.23.00A279
19. Kim K-H et al (2014) Phosphorescent dye-based supramolecules for high-efficiency organic light-emitting diodes. Nat Commun 5

20. Kim K-H, Ahn ES, Huh J-S, Kim Y-H, Kim J-J (2016) Design of heteroleptic Ir complexes with horizontal emitting dipoles for highly efficient organic light-emitting diodes with an external quantum efficiency of 38%. Chem Mater 28:7505–7510
21. Jurow MJ et al (2016) Understanding and predicting the orientation of heteroleptic phosphors in organic light-emitting materials. Nat Mater 15:85–91. https://doi.org/10.1038/nmat4428
22. Kim K-H et al (2015) Controlling emitting dipole orientation with methyl substituents on main ligand of Iridium complexes for highly efficient phosphorescent organic light-emitting diodes. Adv Opt Mater 3:1191–1196. https://doi.org/10.1002/adom.201500141
23. Graf A et al (2014) Correlating the transition dipole moment orientation of phosphorescent emitter molecules in OLEDs with basic material properties. J Mater Chem C 2:10298–10304. https://doi.org/10.1039/c4tc00997e
24. Moon C-K, Kim K-H, Lee JW, Kim J-J (2015) Influence of host molecules on emitting dipole orientation of phosphorescent Iridium complexes. Chem Mater 27:2767–2769. https://doi.org/10.1021/acs.chemmater.5b00469
25. Lampe T et al (2016) Dependence of phosphorescent emitter orientation on deposition technique in doped organic films. Chem Mater 28:712–715. https://doi.org/10.1021/acs.chemmater.5b04607
26. Materials science suite 2016-2, Schrödinger, LLC, New York, NY (2016). https://www.schrodinger.com/citations
27. Banks JL et al (2005) Integrated modeling program, applied chemical theory (IMPACT). J Comput Chem 26:1752–1780
28. Xiang H-F, Xu Z-X, Roy V, Che C-M, Lai P (2007) Method for measurement of the density of thin films of small organic molecules. Rev Sci Instrum 78:034104
29. Gupta J, Nunes C, Vyas S, Jonnalagadda S (2011) Prediction of solubility parameters and miscibility of pharmaceutical compounds by molecular dynamics simulations. J Phys Chem B 115:2014–2023. https://doi.org/10.1021/jp108540n
30. Főrster T (1959) 10th Spiers memorial lecture. Transfer mechanisms of electronic excitation. Discuss Faraday Soc 27:7–17. https://doi.org/10.1039/df9592700007
31. Jaguar 9.2, Schrödinger, LLC, New York, NY (2016)
32. Desmond Molecular Dynamics System 4.6, D. E. Shaw Research, New York, NY (2016); Maestro-Desmond Interoperability Tools, Schrödinger, New York, NY (2016)
33. Bowers KJ et al (2006) Scalable algorithms for molecular dynamics simulations on commodity clusters, SC '06. Proceedings of the 2006 ACM/IEEE Conference on Supercomputing, Tampa, FL, 2006, pp 43. https://doi.org/10.1109/SC.2006.54

Chapter 4
Analysis of the Electronic Structure and Emission Process of Exciplex in Solids

4.1 Introduction

The excited charge-transfer (CT) complex (exciplex) has attracted much attention from researchers in the fields of OLEDs and OPVs. Upon excitation, intermolecular CT between donor (D) and acceptor (A) molecules forms a CT state at a heterojunction. The CT state then either dissociates into positive and negative polarons or relaxes into the ground state by emitting light. These are the main mechanisms governing OPVs [1, 2] and exciplex OLEDs [3, 4], respectively. The bound charges in the CT state separate into free charges when an external electric field is applied. Their binding energy is estimated as ~250 meV [5, 6]. The formation of stable excited CT states without charge separation or endothermic energy transfer [7] results in luminescence with a featureless, red-shifted emission spectrum, and a slow radiative decay rate compared with the D and A molecules. Recently, exciplexes in OLEDs have been actively studied in the context of harvesting triplet excitons by thermally activated delayed fluorescence (TADF) [8–10]. The spatially separated highest occupied molecular orbital (HOMO) and lowest unoccupied molecular orbital (LUMO) in the CT state create a very small energy gap between the singlet and triplet states (ΔE_{ST}). Hence, up-conversion of the triplet state to the singlet state induces TADF [9–11]. Over the past few years, the electroluminescence efficiency of exciplex OLEDs has increased far beyond that of conventional fluorescent OLEDs [8–10, 12–14]. In addition, the exciplex employed as a host not only increases the efficiency of fluorescent OLEDs [15–17] but also helps to reduce the efficiency roll-off of TADF OLEDs, as they convert triplet dopant excitons into singlet dopant excitons [18].

As exciplexes are an important feature of a range of practical devices, many researchers have attempted to understand their characteristics. The energy level alignments between the D and A have a fundamental influence on the photophysical properties of the exciplex such as the emission energy [3, 19], radiative decay rate

C.-K. Moon, *Molecular Orientation and Emission Characteristics of Ir Complexes and Exciplex in Organic Thin Films*, Springer Theses,
https://doi.org/10.1007/978-981-13-6055-8_4

[20], PL quantum yield [13], and reverse intersystem crossing (RISC) yield [8]. The exciplex state is stabilized by the Coulomb attraction between the positively charged D and the negatively charged A, which are separated by a short distance. Therefore, the intermolecular distance between the D and A molecules and their relative orientations play important roles in governing the formation and energy state of the exciplex [21, 22]. In contrast to the well-established microscopic picture at the molecular level, we lack a comprehensive understanding of the properties of exciplexes in solids, despite their use as thin films in organic semiconductor devices. Molecules in solid states are close together and their motions are restricted. Hence, there are many different D-A pairs capable of participating in exciplex formation in geometries that are not the most energetically favorable. The superposition of individual exciplex emissions causes film emission, which broadens the energy levels, kinetic rate constants, and other properties [10, 14]. Therefore, our characterization of the exciplex must take into account the local dimer geometries and the intermolecular CT in the blend. Thus, further investigation of energy levels, ΔE_{ST}, and the emission process of exciplexes is required.

In this chapter, we characterized the configuration and energy state of the D-A heterodimers in a solid-state blend, both experimentally and numerically. Based on our results, we deduced the electronic structures and emission processes of the exciplex. First, we analyzed the wavelength-dependent transient PL decay and emitting dipole orientation (EDO) caused by different local coupling geometries of the dimer in the solid state. Exciplexes that emit relatively short wavelengths were found to be aligned side-by-side and have relatively small CTs from the D to the A molecule, and vice versa. Next, we modeled an amorphous blend using a MD simulation and modeled the exciplex energies and emission in the film. The CT, singlet, and triplet states were found to depend on the intermolecular distance between the D and A molecules and their relative orientations. This is consistent with our experimental observations. Finally, based on the experimental observations, quantum chemical calculations, and MD simulations, we deduced the electronic structure of solid-state exciplexes.

4.2 Results and Discussion

We formed an exciplex layer by co-depositing a 1:1 molar ratio tris(4-carbazoyl-9-ylphenyl)amine (TCTA) and 4,6-bis(3,5-di(pyridin-4-yl)phenyl)-2-methylpyrimidine (B4PyMPM) onto a fused silica substrate. The chemical structures and energy levels of TCTA and B4PyMPM are shown in Fig. 4.1a. The energy levels of TCTA and B4PyMPM were taken from literature. The HOMO levels were measured by ultraviolet photoelectron spectroscopy and the LUMO levels were calculated from the HOMO levels and the absorption edge of the UV-vis absorption spectra, respectively [23, 24].

The absorption spectrum of the film fitted to a linear combination of the absorption spectra of TCTA and B4PyMPM indicated that no aggregation or formation of

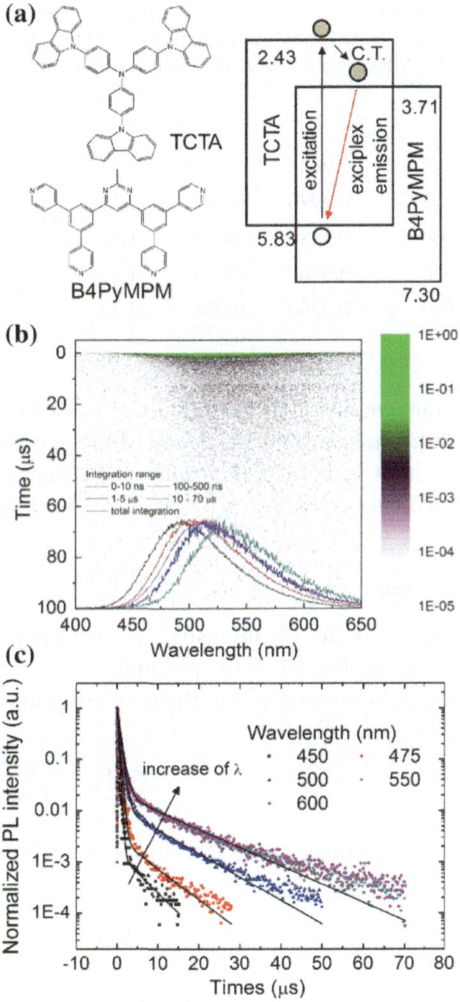

Fig. 4.1 **a** Chemical structures of tris(4-carbazoyl-9-ylphenyl)amine (TCTA) and 4,6-bis(3,5-di(pyridin-4-yl)phenyl)-2-methylpyrimidine (B4PyMPM), with energy levels. Exciplex emission is caused by excitation followed by charge transfer (CT), which results in dimeric electron transitions. **b** A streak image and time-resolved photoluminescence (PL) spectra of the exciplex-es. **c** Wavelength-dependent transient PL decay and fit lines

the CT complex had occurred in the ground state [15]. Figure 4.1b shows a streak image and time-resolved emission spectra of the exciplex. These were measured using a streak camera (Hamamatsu Photonics) and a N_2 pulse laser (337 nm, Usho Optical Systems Co., Ltd). The green and purple colors indicate prompt and delayed exciplex fluorescence, respectively. The integrated emission spectrum is dominated by exciplex emission. A gradual spectral red-shift was observed, with the emission peak shift from $\lambda = 493$ nm at the beginning of the prompt emission region (t = 0–10 ns) to $\lambda = 535$ nm at the end of the delayed emission region (t = 10–70 μs). This continuous spectral red-shift over the delay time is an interesting feature of the solid-state exciplex, and is thought to be caused by a polarization effect in the host medium [9], the broad distribution of the energy level from the different geometric

Table 4.1 Prefactors (A and B), kinetic constants (k_p, k_d, and $k_{ISC}k_{RISC}$), and the proportion of delayed emission (Γ) to total emission in transient photoluminescence decay curves of the TCTA: B4PyMPM exciplexes at different wavelengths

λ (nm)	A	B	k_p ($\times 10^6$ s^{-1})	k_d ($\times 10^4$ s^{-1})	Γ	$k_{ISC}k_{RISC}$ ($\times 10^{10}$ s^{-1})
450	0.9985	0.0015	3.13	17.8	0.026	1.46
475	0.9970	0.003	2.33	13.9	0.048	1.62
500	0.9880	0.012	1.54	10.5	0.151	2.84
550	0.9760	0.024	1.33	7.41	0.307	4.27
600	0.9760	0.024	1.33	7.41	0.307	4.27

arrangements [10], different degrees of CT [25], and inchworm-like CT diffusion [26]. The transient PL decay of the exciplex depends on the emission wavelength. Figure 4.1c depicts the transient PL curves at different wavelengths along with the fitted curves using

$$I = A\ \exp(-k_p t) + B\ \exp(-k_d t), \tag{4.1}$$

where I is the PL intensity, $k_{p,d}$ are the prompt/delayed decay rates, and A and B are the prefactors of the prompt/delayed decays, respectively. The proportion of the delayed emission (Γ) to the total emission is

$$\Gamma = \frac{B/k_d}{A/k_p + B/k_d}. \tag{4.2}$$

The product of the ISC and RISC rates can be derived as follows:

$$k_{ISC}k_{RISC} = \frac{(k_p)^2}{A/B + 1}. \tag{4.3}$$

The values of A, B, k_p, k_d, Γ, and $k_{ISC}k_{RISC}$ of the exciplex at different wavelengths are summarized in Table 4.1. As the wavelength increases, both k_p and k_d decrease but the delayed emission (Γ) and the product of the ISC and RISC rates ($k_{ISC}k_{RISC}$) increase. For instance, Γ is only 3% at 450 nm, but is more than 30% at wavelengths longer than 500 nm, indicating the increase of the CT character at long wavelengths. The increased Γ and $k_{ISC}k_{RISC}$ at longer wavelengths indicate that the lower-energy exciplex has larger CT characters than the high-energy exciplexes. Along with that, the decreased k_p and k_d imply smaller oscillator strength of the low energy exciplex.

We investigated the EDO of the exciplex in films using angle-dependent PL analysis. Figure 4.2a shows the angle-dependent p-polarized PL [27] profiles of the vacuum deposited TCTA:B4PyMPM film at different wavelengths, along with EDO fit lines, which we calculated using an optical simulation [28]. Interestingly, the EDO depends on the emission wavelength. The vertical component of the EDO gradually

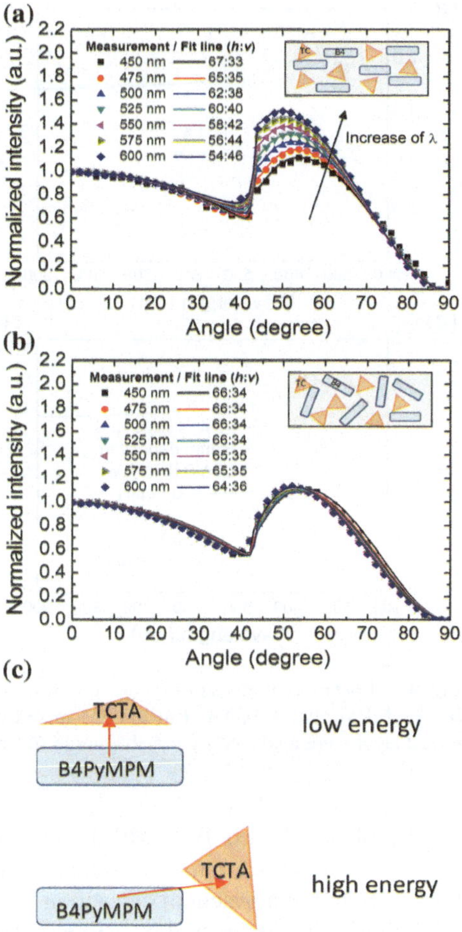

Fig. 4.2 Angle-dependent PL patterns of TCTA:B4PyMPM exciplex layers prepared by **a** thermal evaporation and **b** spin coating. The B4PyMPM and TCTA molecules are depicted as rectangles and triangles, respectively. B4PyMPM molecules have in-plane molecular alignments in the thermally evaporated layer but are ran-domly oriented in the spin-coated layer. The emitting dipole orientation (EDO) of the thermally evaporated exciplex layer shifts gradually towards the vertical direction as the emission wavelength increases; however, the EDO of the exciplex in the spin-coated layer is nearly random. **c** Schematic diagrams illustrating the relationship between exciplex energies and dimer configurations for the co-facial alignment (top) and the side-by-side alignment (bottom)

increases from random (the horizontal to vertical components ratio h:v = 67:33) to a preferred vertical orientation (h:v = 54:46) as the wavelength increases from 450 to 600 nm. In contrast, the spin-coated TCTA:B4PyMPM film exhibited random EDOs and little wavelength dependence, as depicted in Fig. 4.2b.

The key difference between these two cases can be understood from the different orientation of B4PyMPM molecules in the films manifested by the refractive indices of the films. Refractive indices of vacuum-deposited TCTA, B4PyMPM, and TCTA:B4PyMPM films, and a spin coated TCTA:B4PyMPM film are shown in Fig. 4.3 The vacuum-deposited pristine B4PyMPM and mixed films display strong birefringence in contrast to the isotropic refractive indices of the solution-cast mixed film, indicating that the B4PyMPM molecules have preferred in-plain orientations in the vacuum-deposited films, but random orientations in the solution-cast film.

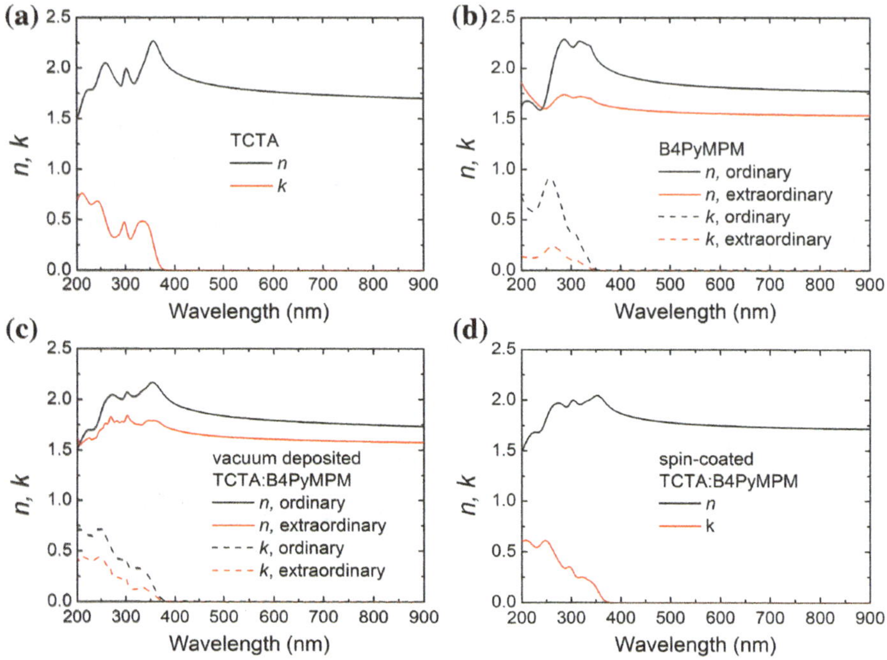

Fig. 4.3 Refractive indices of organic thin films. Optical constants of **a** vacuum deposited TCTA, **b** B4PyMPM, and **c** TCTA: B4PMYPM mixed layers, and **d** a spin-coated TCTA: B4PyMPM mixed layer were analyzed by variable angle spectroscopy ellipsometry

The planar-shaped B4PyMPM molecule has strong intermolecular hydrogen bonds. These induce parallel orientation of the molecular planes with respect to the substrate during vacuum deposition [23]. Considering the in-plane orientation of B4PMYPM molecules in the vacuum-deposited mixed film and the intermolecular electronic transition from TCTA to the B4PyMPM molecules, the vertical EDO can be obtained from the co-facial alignment of TCTA to the B4PyMPM molecules and the horizontal EDO from the side-by-side alignment, as schematically depicted in Fig. 4.2c.

All of the results shown in Fig. 4.2 and the above consideration correlate exciplex geometries with the emission characteristics. The co-facial alignment of TCTA and B4PyMPM molecules in the vacuum-deposited film (inset of Fig. 4.2a) results in a higher vertical EDO, shorter distances between the molecules, longer emission wavelength (lower exciplex energy), slower decay rate, and a larger proportion of the delayed component contributing to the transient PL than the side-by-side alignment. In other words, the wavelength-dependent emission characteristics originate from differences in the geometric arrangements of D and A in the mixed layer. The continuous red-shift in transient PL can be considered as a superposition of fast-decaying, high-energy exciplexes and slow-decaying, low-energy exciplexes. The different emission characters depending on the emission energies are related to the coupling geometries and the degree of CT between D and A molecules in the blend.

The exciplex emission in a blend can be explained by a superposition of the CT emission from D-A pairs. The probability of the CT and the CT emission energy depends on the intermolecular distance of D and A. Therefore, the energy distribution of exciplexes in a blend is expressed by the following equation:

$$\rho(r) = n(r) \cdot \delta(r), \tag{4.4}$$

where ρ and n are the number of exciplexes and the number of D-A pairs with intercharge distance between D(+) and A(-), r, respectively, and δ is the probability of the intermolecular CT at distance r. Since the probability of the CT decreases exponentially with the distance due to the decrease of electron coupling between D and A, we describe δ based on Marcus theory as [29]

$$\delta(r) = \delta_0 \exp\{-\beta(r - r_0)\}, \tag{4.5}$$

where δ_0 is the prefactor, r_0 is the van der Waals contact distance, and β is a constant that depends on the environment. The exciplex energy can be expressed by considering the ionization potential of the donor (IP_D), the electron affinity of the acceptor (EA_A), and the stabilization energy (C) as follows [19]:

$$E = IP_D - EA_A - C. \tag{4.6}$$

The stabilization energy includes the electron-lattice coupling, the polarization effect [30], the electrostatic attraction between the charged molecules [19] and so on. The exciplex energy according to the intercharge distance can be written as follows:

$$E(r) = A - \frac{B}{r}. \tag{4.7}$$

Here, A is considered to be the sum of $IP_D - EA_A$ as well as the stabilization energy independent of the D-A distance and B represents the constant of electrostatic attraction in the exciplex. B is 1 according to Mulliken rule but this is adequate only for the point-to-point charge transfer. It is generally less than 1 because of the nonlocality of charge distribution in molecules and the dielectric screening effect [31]. Now the energy distribution of the exciplex can be written as

$$\rho(E) = \rho[r(E)], \tag{4.8}$$

where $r(E)$ is the inverse function of Eq. (4.7).

MD simulations and quantum chemical calculations were performed to obtain $E(r)$ and $\rho(r)$ A random blend of 500 TCTA and B4PyMPM molecules (1:1 molar ratio) was constructed using an OPLS_2005 force field and an NPT ensemble at 1 atm and 300 K, as shown in Fig. 4.4a. We randomly extracted 483 D-A pairs with

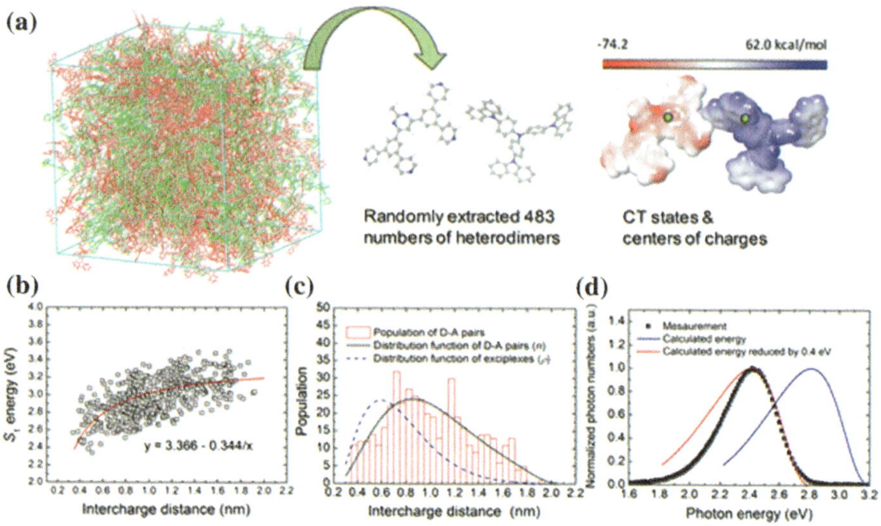

Fig. 4.4 **a** A blend of 250 TCTA and B4PYMPM molecules, formed by a molecular dynamics simulation (left); one random di-mer was extracted from the 40 heterodimers in the blend (middle), and the electrostatic potential surface of a charge-transferred TCTA(+): B4PYMPM(−) dimer was calculated by constrained density functional theory. **b** The calculated CT state energies. The red line indicates an approximated function. **c** Distribution of D-A pairs and CT complexes; the black solid line is a sixth-order polynomial function ($15.57 - 209.8x + 854.9x^2 - 1249x^3 + 856.8x^4 - 286.7x^5 + 37.75x^6$) mimicking the distribution. **d** Comparison between the calculated energy spectrum and the measured emission spectrum

interatomic distances less than 0.4 nm from this blend and calculated the intermolecular CT excitation energies of the dimers using time-dependent density functional theory (TD-DFT) calculations; An example of the charged TCTA(+):B4PyMPM(-) dimer is displayed in Fig. 4.4a. The centers of the charges are located close to each other owing to electrostatic attraction. The distance between centers of charges is taken as the intercharge distance r. Figure 4.4b shows the calculated E against the intercharge distance defined as the distance between the centers of TCTA(+) and B4PyMPM(−), obtained by constrained DFT calculations [32]. The energy decreases with r^{-1}, consistent with Eq. (4.7); thus, the function can be approximated as $E(r) = 3.366 - 0.344/r$.

The distribution of the intercharge distance, $n(r)$, in the 483 D-A pairs is depicted in Fig. 4.4c as red bars. The distribution was approximated by a sixth-order polynomial function (solid line). We employed $\beta = 2.4\ \text{nm}^{-1}$ to calculate $\delta(r)$. Multiplication of the $n(r)$ and $\delta(r)$ results in the distribution of exciplexes ($\rho(r)$, broken line). The distribution of exciplexes indicates that the population reaches a maximum at $r = 0.6$ nm and decreases at longer distances due to the low probability of CT. This agrees with the previous study in which the CT between aromatic amino acids attached to polypeptides was hardly observed at distances longer than 1.2 nm [33]. Figure 4.4d shows the calculated emission spectrum in excellent agreement with the

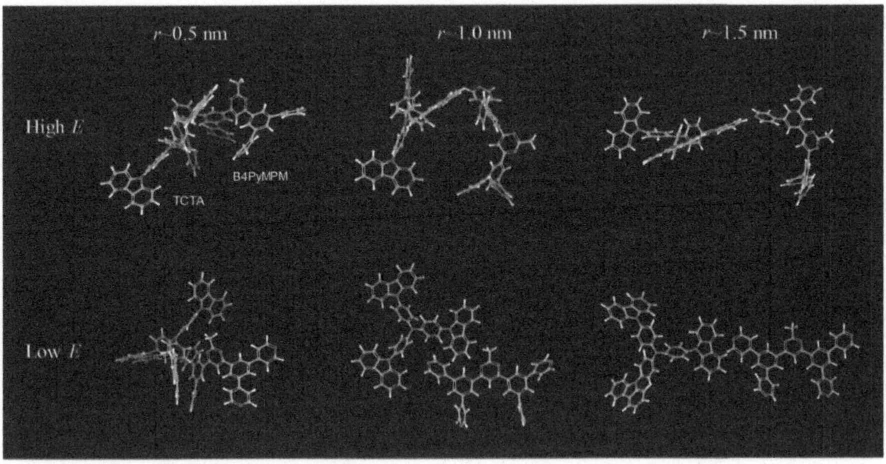

Fig. 4.5 Geometries of TCTA and B4PyMPM

experimental one for the selected value $\beta = 2.4\ \text{nm}^{-1}$. In the figure, $E(r)$ values were reduced by 0.40 eV for the fitting. The discrepancy in energy values may come from the calculation method that considers CT excitation in the fixed dimer geometry. Vibrational relaxation would reduce the energy before the emission.

The calculated S_1 energies have a distribution of ± 0.3 eV compared to the energy–distance trend curve, which can be explained by different molecular shapes in the solid. B4PyMPM prefers a planar molecular structure in the ground state [23]; however, its geometry cannot be ideal in the solid. When a planar structure is maintained, the transition energy is small and the energy is large when the molecule is twisted or has distortion as show in Fig. 4.5.

Figure 4.6a illustrates the dimer arrangements with natural transition orbitals (NTOs) [32] for two different S_1 states. Holes (green and purple surfaces) and particles (red and blue surfaces) are localized on TCTA and B4PyMPM molecules, respectively, indicating CT transitions. The disk-like molecular shapes of TCTA and B4PyMPM result in a longer intermolecular (intercharge) distance of the dimer aligned side-by-side compared with that aligned face-to-face. Therefore, the transition energy of the former dimer (3.01 eV, calculated value) is larger than that of the latter dimer (2.38 eV, calculated value). This is consistent with the gradual shift and the energies according to the dimer alignment that we have described in Fig. 4.2a. The relationship of the singlet-triplet transition energy of the exciplex is depicted in Fig. 4.6b. When the singlet energy is less than 2.8 eV, there is little difference between singlet and triplet energies. NTOs for T_1 represent ^{3}CT states. When the singlet energy becomes larger, the triplet energy does not continue to increase, but remains fixed between 2.8 and 3.2 eV depending on the molecular configuration in solids. NTOs for T_1 in this region represent ^{3}LE states of TCTA or B4PyMPM. Owing to the pinning of the T_1 levels, the T_1 band is narrower than the S_1 band as schematically represented in Fig. 4.6c. The S_1 band mostly consists of CT states, and

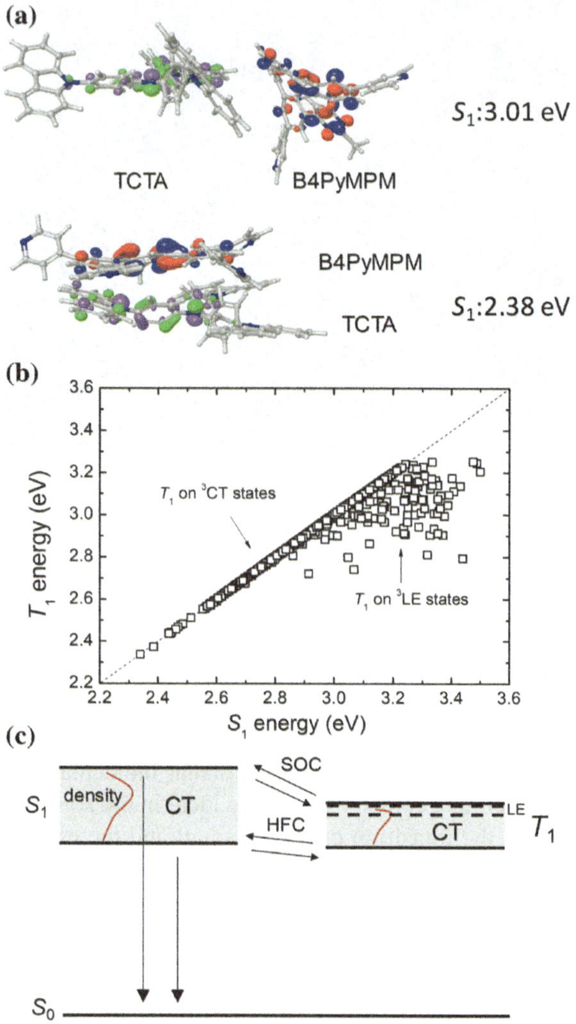

Fig. 4.6 **a** Singlet transition energies depending on dimer arrangements and their natural transition orbital (NTOs) (hole: green and purple sur-faces, particle: red and blue surfaces). **b** Relationship between simulated singlet and triplet transition energies of TCTA:B4PyMPM dimers. **c** Schematic energy band diagram of the exciplex, with broadly distributed energy levels. The high- and low-energy exciplexes have different emission processes and singlet-triplet electronic transition rates

the T_1 band consists of LE states at high energy levels and CT states at the lower levels than the LE levels.

High energy exciplexes have large S_1-T_1 gaps corresponding to the ^{1}CT–^{3}LE gaps and low energy exciplexes have small S_1-T_1 gaps corresponding to the ^{1}CT–^{3}CT gaps.

The increased delayed fluorescence with deceasing energy can be understood based on the electronic structure in in Fig. 4.4c. The ISC and RISC of the exciplex are the processes induced by hyperfine coupling (HFC) between ^{1}CT and ^{3}CT states [34] and spin-orbit coupling (SOC) between ^{1}CT and ^{3}LE states [35–39]. High-energy exciplexes possessing large ^{1}CT–^{3}LE gaps show, small proportion of delayed fluorescence (Table 4.1) due to small SOC. With decreasing exciplex energy, the SOC rate increases due to decreasing the ^{1}CT–^{3}LE gap to increase the delayed fluorescence until $S_1 = {}^1CT \approx {}^3LE \approx {}^3CT$. Further decrease of exciplex energy increase the ^{1}CT–^{3}LE gap to reduce the SOC. In contrast, the HFC increases due to the reduction of the ^{1}CT–^{3}CT gap, Therefore, HFC is regarded as the RISC path in the exciplex when ^{3}CT is located below ^{3}LE level. When the exciplex energy is lower than 2.25 eV (550 nm of the emission wavelength, Table 4.1) corresponding to 2.65 eV in the calculation, the proportion of the delayed fluorescence no longer decreases, which can be attributed to the constant HFC rate between ^{1}CT and ^{3}CT states. We speculate that the fast prompt/delayed fluorescence rates in high-energy exciplex is by mixing ^{1}CT and ^{1}LE states for S_1, resulting in the increased oscillator strength.

In summary, the solid-state exciplex has a broad energy band due to the various dimer geometries, which have different distances between D and A molecules. Exciplexes with shorter D-A distances have lower energy due to the Coulomb interaction proportional to r^{-1}. Co-facially aligned heterodimers have a shorter intermolecular distance and form lower energy exciplexes than side-by-side aligned dimers. If the molecules are horizontally oriented in the blended film, the EDO gradually shifts to the vertical direction as the exciplex energy decreases. Coexistence of ^{3}CT and ^{3}LE states results in the different kinetics of the exciplex. The high-energy exciplex has a large ΔE_{ST} with slow electronic transition rates and a reduction in delayed fluorescence, which is expressed as a narrower triplet band than the singlet band. The spectral-red shift with time delay originates from the superposition of fast-decaying, high-energy exciplexes and the slow-decaying, low-energy exciplexes. In addition, these results clearly show that the formation and emission properties of the excited CT state are influenced by the molecular structure, D-A distance, molecular orientation, energy level alignments, CT rate, D/A ratio, etc., of the blends. We expect our diagram of the electronic structure of excited CT states to help researchers to understand various properties observed in OPVs and exciplex OLEDs in which excited CT states play an important role.

References

1. Yu G, Gao J, Hummelen JC, Wudl F, Heeger AJ (1995) Polymer photovoltiac cells: Enhanced efficiencies via a network of internal donor-acceptor heterojunctions. Science 270:1789
2. Sariciftci NS, Smilowitz L, Heeger AJ, Wudl F (1992) Photoinduced electron transfer from a conducting polymer to buckminsterfullerene. Science 258(5087):1474–1476
3. Jenekhe SA, Osaheni JA (1994) Excimers and exciplexes of conjugated polymers. Science 265(5173):765–768
4. Wang JF et al (1998) Exciplex electroluminescence from organic bilayer devices composed of triphenyldiamine and quinoxaline derivatives. Adv Mater 10:230–233
5. Gélinas S et al (2011) The binding energy of charge-transfer excitons localized at polymeric semiconductor heterojunctions. J Phys Chem C 115:7114–7119

6. Offermans T, van Hal PA, Meskers SC, Koetse MM, Janssen RA (2005) Exciplex dynamics in a blend of π-conjugated polymers with electron donating and accepting properties: MDMO-PPV and PCNEPV. Phys Rev B 72:045213
7. Morteani AC, Friend RH, Silva C (2004) Endothermic exciplex–exciton energy-transfer in a blue-emitting polymeric heterojunction system. Chem Phys Lett 391:81–84
8. Goushi K, Adachi C (2012) Efficient organic light-emitting diodes through up-conversion from triplet to singlet excited states of exciplexes. Appl Phys Lett 101:023306. https://doi.org/10.1063/1.4737006
9. Goushi K, Yoshida K, Sato K, Adachi C (2012) Organic light-emitting diodes employing efficient reverse intersystem crossing for triplet-to-singlet state conversion. Nat Photonics 6:253–258. https://doi.org/10.1038/nphoton.2012.31
10. Park Y-S, Kim K-H, Kim J-J (2013) Efficient triplet harvesting by fluorescent molecules through exciplexes for high efficiency organic light-emitting diodes. Appl Phys Lett 102:153306. https://doi.org/10.1063/1.4802716
11. dos Santos PL, Dias FB, Monkman AP (2016) Investigation of the mechanisms giving rise to TADF in exciplex states. J Phys Chem C 120:18259–18267
12. Liu XK et al (2015) Remanagement of singlet and triplet excitons in single-emissive-layer hybrid white organic light-emitting devices using thermally activated delayed fluorescent blue exciplex. Adv Mater 27:7079–7085. https://doi.org/10.1002/adma.201502897
13. Liu XK et al (2015) Prediction and design of efficient exciplex emitters for high-efficiency, thermally activated delayed-fluorescence organic light-emitting diodes. Adv Mater 27:2378–2383. https://doi.org/10.1002/adma.201405062
14. Kim K-H, Yoo S-J, Kim J-J (2016) Boosting triplet harvest by reducing nonradiative transition of exciplex toward fluorescent organic light-emitting diodes with 100% internal quantum efficiency. Chem Mater 28:1936–1941
15. Kim KH, Moon CK, Sun JW, Sim B, Kim JJ (2015) Triplet harvesting by a conventional fluorescent emitter using reverse intersystem crossing of host triplet exciplex. Adv Optical Mater 3:895–899
16. Zhao B et al (2015) Highly efficient red OLEDs using DCJTB as the dopant and delayed fluorescent exciplex as the host. Sci Rep 5:10697
17. Kim HG, Kim KH, Moon CK, Kim JJ (2017) Harnessing triplet excited states by fluorescent dopant utilizing codoped phosphorescent dopant in exciplex host for efficient fluorescent organic light emitting diodes. Adv Optical Mater 5(3):1600749
18. Moon CK et al (2017) Combined inter-and intramolecular charge-transfer processes for highly efficient fluorescent organic light-emitting diodes with reduced triplet exciton quenching. Adv Mater 29(17):1606448. https://doi.org/10.1002/adma.201606448
19. Müllen K, Scherf U (2006) Organic light emitting devices: synthesis, properties and applications. Wiley
20. Gould IR, Young RH, Mueller LJ, Albrecht A, Farid S (1994) Electronic structures of exciplexes and excited charge-transfer complexes. J Am Chem Soc 116:8188–8199
21. Sreearunothai P et al (2006) Influence of copolymer interface orientation on the optical emission of polymeric semiconductor heterojunctions. Phys Rev Lett 96:117403
22. Huang Y-S et al (2008) Electronic structures of interfacial states formed at polymeric semiconductor heterojunctions. Nat Mater 7:483
23. Sasabe H et al (2011) Influence of substituted pyridine rings on physical properties and electron mobilities of 2-methylpyrimidine skeleton-based electron transporters. Adv Func Mater 21:336–342
24. Su SJ, Gonmori E, Sasabe H, Kido J (2008) Highly efficient organic blue-and white-light-emitting devices having a carrier-and exciton-confining structure for reduced efficiency roll-off. Adv Mater 20:4189–4194
25. Graves D, Jankus V, Dias FB, Monkman A (2014) Photophysical investigation of the thermally activated delayed emission from films of m-MTDATA: PBD exciplex. Adv Func Mater 24:2343–2351. https://doi.org/10.1002/adfm.201303389

26. Deotare PB et al (2015) Nanoscale transport of charge-transfer states in organic donor-acceptor blends. Nat Mater 14:1130–1134. https://doi.org/10.1038/nmat4424
27. Frischeisen J, Yokoyama D, Adachi C, Brütting W (2010) Determination of molecular dipole orientation in doped fluorescent organic thin films by photoluminescence measurements. Appl Phys Lett 96:073302. https://doi.org/10.1063/1.3309705
28. Moon CK, Kim SY, Lee JH, Kim JJ (2015) Luminescence from oriented emitting dipoles in a birefringent medium. Optics Express 23(7):A279–A291. https://doi.org/10.1364/oe.23.00a279
29. Marcus RA, Sutin N (1985) Electron transfers in chemistry and biology. Biochimica et Biophysica Acta (BBA)-Rev Bioenergetics 811:265–322
30. Brédas JL, Cornil J, Heeger AJ (1996) The exciton binding energy in luminescent conjugated polymers. Adv Mater 8:447–452
31. Baumeier B, Andrienko D, Rohlfing M (2012) Frenkel and charge-transfer excitations in donor-acceptor complexes from many-body green's functions theory. J Chem Theory Comput 8:2790–2795. https://doi.org/10.1021/ct300311x
32. Martin RL (2003) Natural transition orbitals. J Chem Phys 118:4775–4777. https://doi.org/10.1063/1.1558471
33. Inai Y, Sisido M, Imanishi Y (1990) Distance and orientation dependence of electron transfer and exciplex formation of naphthyl and p-dimethylanilino groups fixed on a helical polypeptide chain. J Phys Chem 94:6237–6243
34. Ogiwara T, Wakikawa Y, Ikoma T (2015) Mechanism of intersystem crossing of thermally activated delayed fluorescence molecules. J Phys Chem A 119:3415–3418
35. Etherington MK, Gibson J, Higginbotham HF, Penfold TJ, Monkman AP (2016) Revealing the spin-vibronic coupling mechanism of thermally activated delayed fluorescence. Nat Commun 7:13680. https://doi.org/10.1038/ncomms13680
36. Samanta PK, Kim D, Coropceanu V, Bredas JL (2017) Up-conversion intersystem crossing rates in organic emitters for thermally activated delayed fluorescence: impact of the nature of singlet versus triplet excited states. J Am Chem Soc 139:4042–4051. https://doi.org/10.1021/jacs.6b12124
37. Gibson J, Monkman AP, Penfold TJ (2016) The importance of vibronic coupling for efficient reverse intersystem crossing in thermally activated delayed fluorescence molecules. Chem Phys Chem 17:2956–2961. https://doi.org/10.1002/cphc.201600662
38. Dias FB et al (2016) The role of local triplet excited states and D-A relative orientation in thermally activated delayed fluorescence: photophysics and devices. Adv Sci 3(12):1600080. https://doi.org/10.1002/advs.201600080
39. Lee K, Kim D (2016) Local-excitation versus charge-transfer characters in the triplet state: theoretical insight into the singlet-triplet energy differences of carbazolyl-phthalonitrile-based thermally activated delayed fluorescence materials. J Phys Chem C 120:28330–28336. https://doi.org/10.1021/acs.jpcc.6b10161

Chapter 5
Summary and Conclusion

This thesis presented the principle and optical effects of the molecular orientation for thin films and OLEDs. It also introduced the application of molecular orientation in the analysis of the electronic structure of the bimolecular excited state.

In Chap. 2, an optical model for emission in a birefringent medium was presented and the theoretical model was validated by applying it to the emission from a thin film and OLEDs. The optical birefringence affects not only far-field radiation properties such as emission spectra from the film and OLEDs but also the efficiency of the OLEDs. The emitting dipole orientation of a heteroleptic Ir complex doped in the birefringent medium was successfully analyzed from the far-field radiation pattern of a thin film. In addition, the birefringent model provided a precise analysis of angle-dependent EL spectra and EQEs of OLEDs.

In Chap. 3, the molecular orientation of Ir complexes in vacuum-deposited organic layers were demonstrated by quantum mechanical analyses and molecular dynamics simulations. Ir complexes have spontaneous alignments by local electrostatic and van der Waals interactions with nearest host molecules interacting in a smaller scale than a molecule. Attractive interactions between the aromatic ligands of Ir complexes and hosts induce the horizontal dipole orientation and the interactions rely on the molecular structures of the dopant and the host. The increase of the dispersion force and electrostatic attraction on the aromatic ligands was effective on increasing the horizontal dipole orientation with lowered non-bonded interaction energy.

In Chap. 4, the exciplex emission was modeled by a quantum chemical characterization of the CT states in a donor-acceptor blend. Various donor-acceptor geometries in the blend resulted in broad distributions of the energy and excited state dynamics. The energy spectrum of the solid-state exciplex could be understood by a product of the exciplex energy, the density of donor-acceptor dimer, and the probability of CT as functions of the CT distance. In addition, the larger delayed fluorescence from the lower-energy exciplex was explained by the narrower energy band of the triplet state than that of the singlet state in the solid state.

C.-K. Moon, *Molecular Orientation and Emission Characteristics of Ir Complexes and Exciplex in Organic Thin Films*, Springer Theses,
https://doi.org/10.1007/978-981-13-6055-8_5

Curriculum Vitae

Name	Chang-Ki Moon
Address	Rm #120, Bldg #39, Research Institute of Advanced Materials, Seoul National University, Seoul, 08826, South Korea
Phone	+82-2-875-2412
Cell phone	+82-10-5100-2439
E-mail	nuenunnue@snu.ac.kr

Education

Ph.D. 2014.3–2017.8	Department of Materials Science and Engineering, Seoul National University.
M.A. 2012.3–2014.2	Department of Materials Science and Engineering, Seoul National University.
B.A. 2008.3–2012.2	Department of Materials Science and Engineering, Seoul National University.

Experiences

2015.6–2016.5	Development and commercialization of OLED optical simulation software and equipment for an analysis of emitting dipole orientation in thin films: "Luxol-OLED simulator" & "Luxol-OLED analyzer".
2015.1–2015.3	Visiting and Collaboration with Max Plank Institute for Polymer research in Mainz (Dr. Denis Andrienko), Germany through International Research Training Group (IRTG) program.

C.-K. Moon, *Molecular Orientation and Emission Characteristics of Ir Complexes and Exciplex in Organic Thin Films*, Springer Theses,
https://doi.org/10.1007/978-981-13-6055-8

Awards

1. "Best Paper Award (Silver)", The 18th International Meeting on Information Display (IMID 2018), August 30, 2018.
2. "Best Paper Award", Schrodinger's 2017 Excellence in Materials Science Applications Publication Contest, Schrodinger Inc. February 15, 2018.
3. "Best Ph.D. Thesis Award in 2017", Department of Materials Science and Engineering, Seoul National University, December 13, 2017.

Research Area and Interest

- Materials theory and computation
- Molecular simulations and DFT calculations
- Organic electronics and photonics
  - Organic semiconducting materials
  - Electrical and optical processes in organic semiconductors
  - Exciton dynamics
  - Inter- and intra-molecular charge transfer and locally excited states
  - Device physics
  - Molecular orientation and arrangement in films
- Optical spectroscopy.

Professional Skills

- Computational skills: Programming (Matlab, Python), quantum chemical calculation (Gaussian09, Jaguar), molecular dynamics simulation (Desmond), film reflection-transmission-absorption simulation, and simulation of luminescence and absorption from thin films, OLEDs, and OPVs.
- Optical spectroscopy: photoluminescence spectrum, absolute photoluminescence quantum efficiency, transient photoluminescence, molecular orientation analysis by angle-dependent photoluminescence spectroscopy, and variable-angle spectroscopic ellipsometry.
- Fabrication and characterization of electroluminescent devices: thermal evaporator, current density-voltage-luminance characteristics, and transient electroluminescence.

The manufacturer's authorised representative in the EU is Springer Nature Customer Service Centre GmbH, Europaplatz 3, 69115 Heidelberg, Germany. If you have any concerns regarding our products, please contact ProductSafety@springernature.com

Printed and bound by CPI Group (UK) Ltd, Croydon, CR0 4YY
30/09/2026
02225465-0004